PROBLEMS FOR MUSICAL ACOUSTICS

William R. Savage

New York

OXFORD UNIVERSITY PRESS

1977

Library of Congress Cataloging in Publication Data

Savage, William R 1926-
Problems for musical acoustics.

Includes index.
1. Music--Acoustics and physics. I. Title.
ML3805.S27 781'.1 77-24934
ISBN 0-19-502251-3

Printed in the United States of America

Acknowledgments

I am most grateful and heavily indebted to Arthur H. Benade in the preparation of this collection of problems. He lent me numerous sets of his class notes, examinations, problems, and sustained my efforts with marginal notes on the draft copies of these problems. I have benefited by a very personal and intensive correspondence course in musical acoustics, that I share with the reader along with my own experience in teaching the "Physics of Sound and Music" here at Iowa.

In addition to the encouragement and assistance provided by Professor Benade, this project has benefited from the helpfulness of a number of my colleagues who have given me access to their class notes, tests, problem sets, and text manuscripts. In alphabetical order according to the institutions in which they teach, these supportive persons include W. Strong, Brigham Young University; J. Traylor, Buena Vista College; R. H. Silsbee, Cornell University; E. Koch, Eastern Illinois University; E. Brock Dale, Kansas State University; D. Zych, New York State University College at Oswego; G. Weinreich, University of Michigan; J. G. Anderson, University of Wisconsin at Eau Claire; and R. Smith, University of Wisconsin at Plattesville.

I cannot forget the technical help in writing given to me by James Wells, of my department staff, and his especially valuable comments on the organization of the material.

Margaret Rich receives thanks for reading the problems and commenting on them from the viewpoint of a former student in my course. John Birkbeck, technical publication illustrator in my department, helped with those illustrations that required photographic processing. Special thanks are owed to Holly Waldron for preparing the draft copies. Very special thanks are owed to Shirley Streeby for attending to details of English usage and for preparing the final copy.

Contents

1

PRELIMINARIES

Problem solving in musical acoustics is historically traceable to some 2500 years ago when Pythagoras worked out relationships between pitch and the length of the vibrating string. Over the intervening centuries a search for scientific understanding has accompanied the musical arts. An inquiring host of natural philosophers, physical scientists, specialists in acoustics, and musicians have developed an immense quantity of information. Out of this body of knowledge have emerged technological applications with significant influences upon the lives of music makers and their audiences. The increasing number of publications and of academic courses in musical acoustics appears to show an expanding and deepening awareness of the relevancies between the art and the science. More and more performers and listeners seem to be striving to follow the progression through the acoustical environment from creation through reception to perception. The recent development of electronic music and the continuous refinement of high-fidelity audio equipment have accentuated the trend toward more awareness and understanding of acoustical phenomena. Not surprisingly, the increased technology in electronics has stimulated interest in historical and traditional music as well as the instruments used in performance.

Problems for Musical Acoustics began as a project suggested by Arthur H. Benade. In undertaking the preparation of a volume of problems to supplement his texts, I trust that I am also providing a convenient collection to add to the instructional wealth of other texts in this field of study.

Admittedly, obviously, this book presents an awesome array of problems. Hopefully they are so arranged that the instructor—and the student—can select what he needs for his purposes. To find the questions suitable for a particular phase of his work, one needs only to look in the index for the key words of the concepts. The material is organized to follow the succession of topics as they appear in Fundamentals of Musical Acoustics by Arthur H. Benade. The chapter headings are the same except that one chapter is devoted to woodwinds and one chapter is devoted to string instruments and the bowed string. The last three chapters of problems are organized to follow the early chapters of The Acoustical Foundations of Music by John Backus. The users of other textbooks and their own class notes should find these problems helpful. Admittedly, these problems do not use mathematics beyond easy multiplication and the use of graphs. Many of these problems are adapted for the use of an electronic hand calculator that has keys for sine, cosine, square root, and logarithmic functions.

Here at the University of Iowa we have a long and fruitful tradition of work in advancing the study of acoustics. This work began in 1856 with the first man to occupy the Chair of Natural Philosophy in the infant university. The Rev. Jared M. Stone considered acoustics to be one of the five principal subjects in his course of study. Instruction—and in later years, research—in the physics of sound continued through the periods of Stone's successors—The Rev. Oliver M. Spencer, Gustavus Hinrichs, Andrew Veblen, etc. Fitting and appropriate perhaps was an item in the Faculty Minutes of the University of Iowa, dated November 19, 1862:

"Prof. Hinrichs to take charge of instrumental music on all occasions at Chapel."

But it was not until George Walter Stewart began his long and productive period (1909-1946) as Professor and Head of the Department of Physics that the University of Iowa became a recognized leader in physical acoustics research and instruction. During most of these years G. W. Stewart conducted and directed a considerable body of research, much of it with useful applications to noise control and in communications.

Contemporary to the work in physical acoustics was the equally distinguished and productive period (1909-1945) of investigation by Carl E. Seashore in the psychological and perceptual aspects of sound. In 1910 Seashore founded a program in the psychology of music. He contributed to instrumentation and measurement in audiology, to investigations on the perception of sounds relating to tone quality, pitch, and vibrato. In an account of his contributions to research in _Science_, Vol. 111, pp. 713-719, 1950, a reference is made to Seashore's view of the importance of direct experience in our understanding. He felt that the hypothetical solution to problems required a test where, for the correlated project, "experimentation (in the laboratory with 'brass instruments' if appropriate) was a prerequisite."

My own interest in this project results from the course I have taught on the Physics of Sound and Music. The conferences here at Iowa on the "Teaching of Acoustics" and the "Physics of Sound and Music" in 1974 and 1976 organized by Bill Hartmann of Michigan State University and myself have also served as motivation for this book of problems for musical acoustics.

These problems have been written so as to distinguish between the physical aspects and the perceived aspects of sound. In working out the problems it may be necessary to consider this distinction with care. For example, if given a few facts about the speed of sound and the length and shape of an air column one can calculate the frequency of the fifth mode (even including end corrections). But to find the pitch of this air column is a really tough task. The pitch depends on the way sounds are produced by a real air column and the way the brain organizes and processes signals originating in the ear. These problems and the material in _Fundamentals of Musical Acoustics_ are designed to emphasize conclusions based on direct observation. The student who conducts the experiments himself will derive the maximum benefit. Instructors who use an open laboratory or provide equipment for informal experiments report high student interest and participation. Students and teachers whose courses are based on the more traditional pedagogical method of beginning with basic definitions, model situations, and then to application should find these problems useful for supplemental study.

2

IMPULSIVE SOUNDS, ALONE AND IN SEQUENCE

2-1. What is the repetition rate for each sequence of pulses shown in the diagram below?

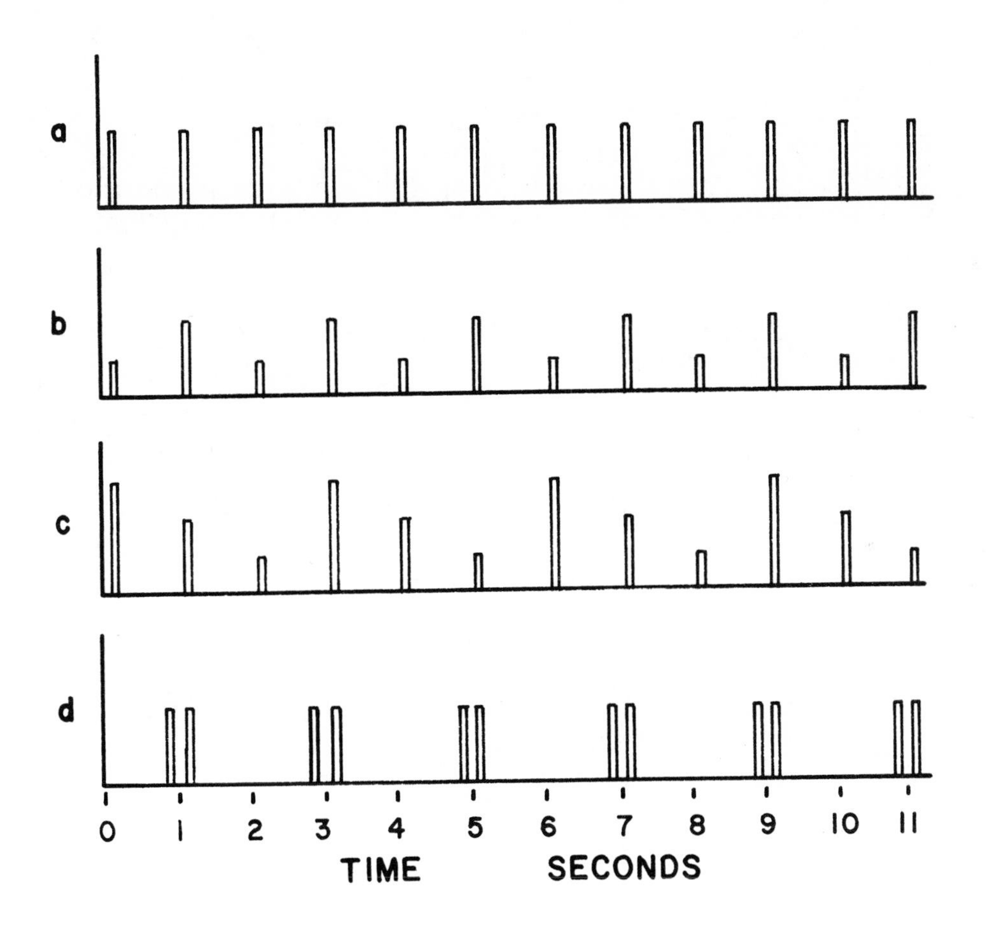

2-2. From Figure 2.1 determine the repetition rate required for a series of pulses to agree with the pitch of the musical note E_4.

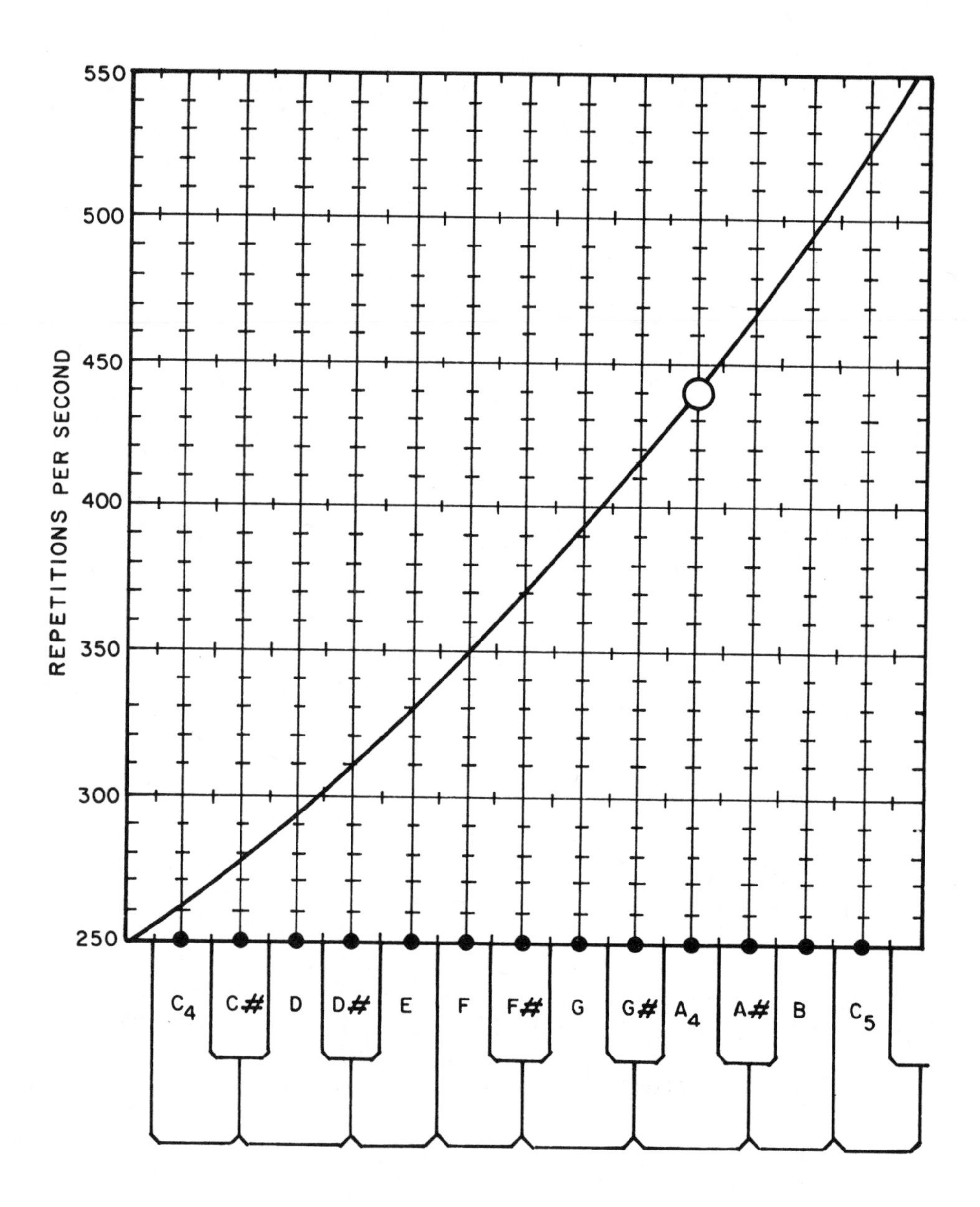

Figure 2.1. A Reference Scale Relating Repetition Rate and Musical Note Name (Adapted from Fundamentals of Musical Acoustics, Arthur H. Benade, Oxford University Press, 1976 with permission.)

2-3. From Figure 2.1 determine the name of the standard musical note whose repetition rate is closest to 400 repetitions per second. What should the exact repetition rate be for this musical note?

2-4. What is the time between pulses for these repetition rates: 20/sec; 100/sec; 120/sec; 256/sec; 440/sec; 1000/sec; 10,000/sec?

2-5. Make graph sequences of impulses that have an interpulse time of 0.010 sec, 0.050 sec, 0.25 sec, and 1.0 sec. Instead of making separate graphs for each sequence, the impulses can be graphed also on a single time axis. What is the repetition rate of the composite sequence?

2-6. What is the repetition rate of the sequence of impulses sketched below? Which musical note would best describe the rate you found for the sequence as determined by comparison to a frequency table for the equal temperament scale shown in Figure 2.1?

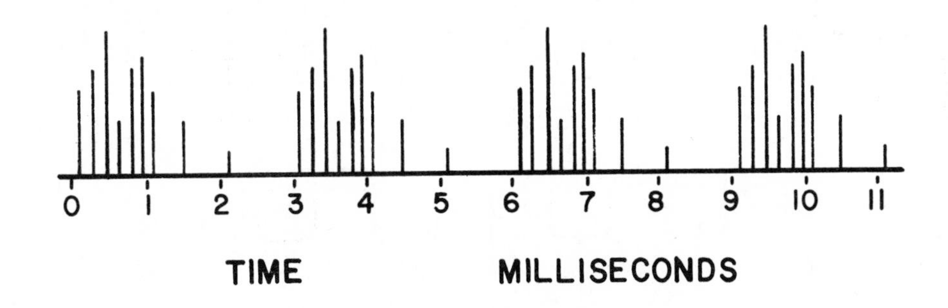

2-7. A child runs along a picket fence while holding a stick so as to tap each picket as it is passed. What is the repetition rate of the taps when 144 pickets are struck uniformly in 1 min? What is the time between taps?

2-8. A person observes two flashing yellow construction warning lights. They are located close enough together so that the flashing of each can be compared with the other as a function of time. Observation of the two lights reveals that they both flash at the same time only once each minute. One of the lights flashes five times per minute and the other six times per minute. Make a pattern diagram that illustrates the pulsations by plotting separate symbols for each light on a time axis which extends for at least 3 min. Discuss the difference between the patterns formed for two different pulses repeating at the same rate and two sources of identical pulses repeating at different rates.

2-9. The diagram shown below represents the possible sequence of pulses for a slightly irregular process such as bubbling or gurgling. Search for possible repetitions of similar pulses, and identify the repetition rate for each. It may be helpful to make a special scale at the edge of a piece of paper that you can slide alongside the pulses in the search for repetitions.

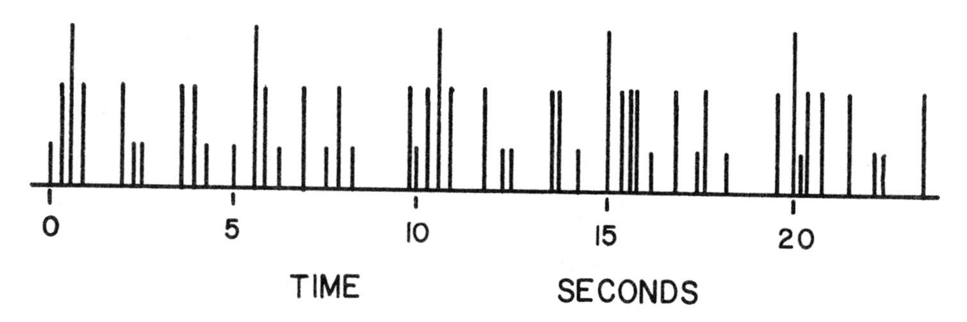

2-10. A very basic interval or spacing between tones is the musical relationship of the octave. Find the nearest note names and repetition rates assigned to the following intervals, showing your calculations: (a) up from C_4 by one octave, (b) up from A_2 by two octaves, (c) down from C_3 by one octave, (d) down from A_4 by one half octave, (e) up from 350 rep/sec by a half octave, and (f) down from 560 rep/sec by one octave.

3

Simple Relations of Sounds and Motions

3-1. Sounds produced by tapping on different objects have different characteristics. Consider the impulse sounds produced by tapping on various objects, such as glasses, bottles, metal lamp shades, paper cups, serving trays, hub caps, and similar objects that you can find. Characterize these sounds with terms that describe their duration. Imagine an oscilloscope trace of the sound, and suggest how the alternation rate of fine-grained reversals might appear. Alternatively, a similar description of the sounds can be made by such descriptors as noisy, bell-like, ringing, musical, thud, clang, etc. So that a comparison of the sounds from different objects can be made, use a single tapper for each series of comparisons. Compare the results of the series of sounds obtained with a hard tapper such as the plastic butt end of a ball-point pen to a series obtained with a soft tapper such as a rubber eraser tip.

3-2. The sounds produced by tapping on an object in different places can have different durations and fine-grained reversals. For this study find an object with a certain amount of symmetry, such as a metal bowl, metal lamp shades, serving trays, dinner plates, brake drums (removed from the car), metal boxes, etc. Characterize or rank the sounds produced by tapping on it with respect to both the location of the tap and to different hammers applied to the same location.

3-3. If you have access to a musical instrument you can produce interesting sounds by tapping on various portions of the instrument. Discuss the general characteristics of the sounds while resisting efforts to associate these tap sounds with the musical sounds produced by the instrument. Avoid tapping on the strings of such instruments as the guitar, harpsichord, piano, and banjo for the sources of sound in this discussion. These sounds are treated separately in other problems.

3-4. A description of sounds produced by tapping on a single object has been requested in the problems above. Consider sounds produced in some slightly more complicated situations. Be careful since things can become very complicated very quickly. Describe the sound with regard to fine-grained reversals and durations for situations such as closing a door, clapping your hands in a long hallway, a gun shot outdoors, thunder, etc. Restrict the discussion to sounds initiated by an impulse.

3-5. Oscilloscope traces, chart paper records, and graphs can present a visual form of the vibrations or alternations of an object as they change with time. You may wish to practice or review the general properties of some simple graphs. Plot the graphs of the simple functions $y = x$, $y = x^2$, and $y = x^3$ over the region $x = +1$ to $x = -1$. As a help in getting started a few values of x^2 and x^3 are already entered in the table of numbers from which you can plot the graph. After these three graphs are neatly plotted and the appropriate curves drawn through the points, indicate the region of the graph in which you would expect the graphs of $y = x^{0.5}$, $y = x^{1.5}$, and $y = x^{2.5}$ to be located. Do this only for the region lying between $x = 0$ and $x = +1$. What mathematical difficulty might arise from an attempt to

Y →

X	-1.0	-0.8	-0.6	-0.4	-0.2	0.2	0.4	0.6	0.8	1.0
X^2	+1.0				+0.04					
X^3	-1.0				-0.008					

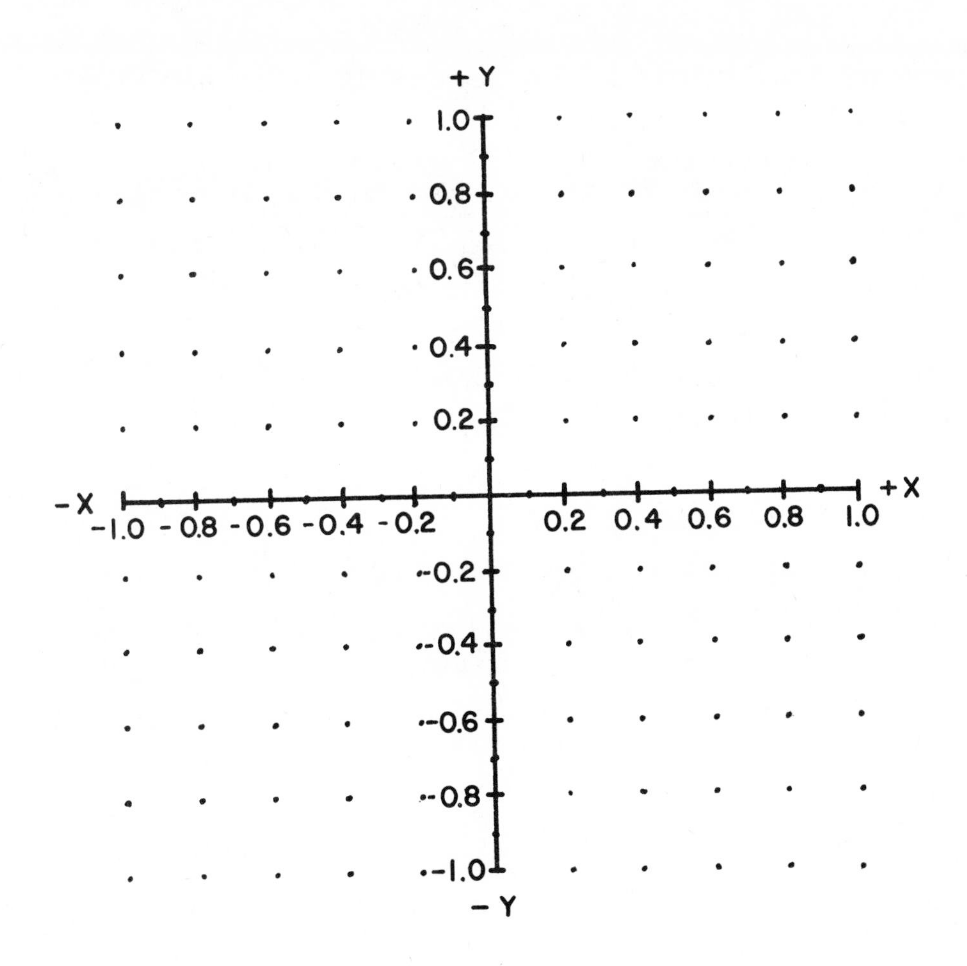

 plot the curve into the region of negative x? Does this mean that curves which are smooth continuations of the positive-x part of the diagram do not exist in this region?

3-6. In Olmsted's 1838 textbook on Natural Philosophy he mentions the observation by Herschel that a glass filled with an effervescent liquid emits a dull sound while the bubbling occurs, which becomes a ringing sound when the fluid becomes clear. This experiment can be repeated by filling a wine glass with warm water. Tap gently and listen to the sound as the bubbles rise to the surface. Describe all the changes you hear from the first taps until the water loses all the bubbles. (A chilled, carbonated liquid is not recommended since a long time is required for the bubbling to cease.)

3-7. One technique for locating studs in houses is to tap the wall with a knuckle and listen for changes in the sound. Experiment with this technique and report the changes in sound that you observe. (You can verify the location of the studs by using a compass or a small magnet suspended from a piece of thread. A commercial stud locator uses a pivoted magnet to find the position of the nails that hold the wallboard to the studs.)

4

Characteristic Frequencies and the Decay of Composite Sounds

4-1. It will be helpful in visualizing the results of combining two vibrations into one composite vibration to combine two graphs into a single summation graph. Use a pair of dividers or a scale to prepare the summation graph of the two disturbances given below.

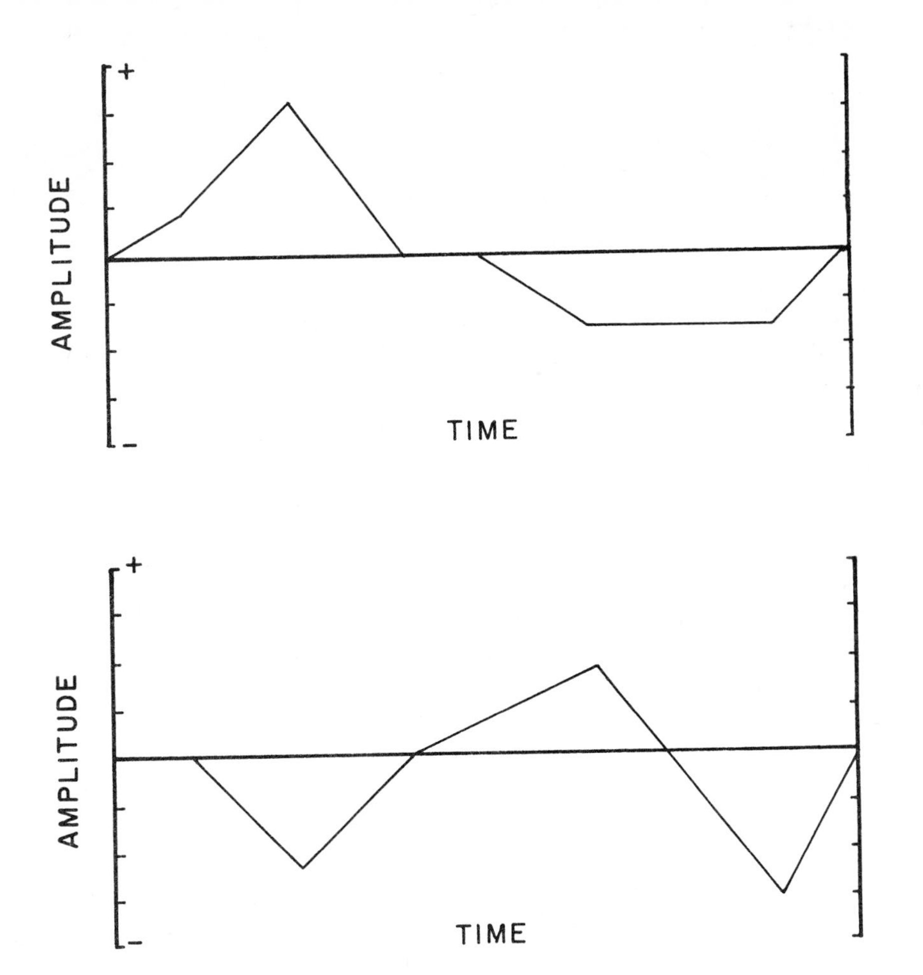

4-2. In order to see an interesting extension of the effect of phase shifts and relative phase in the summation of two disturbances, find the combination (summation) graph of the two sinusoids given below. Use a pair of dividers or a scale to prepare the graph of the disturbance that is the sum of the two disturbances.

a. Equal frequencies with the same initial phase (in phase) and different amplitudes.

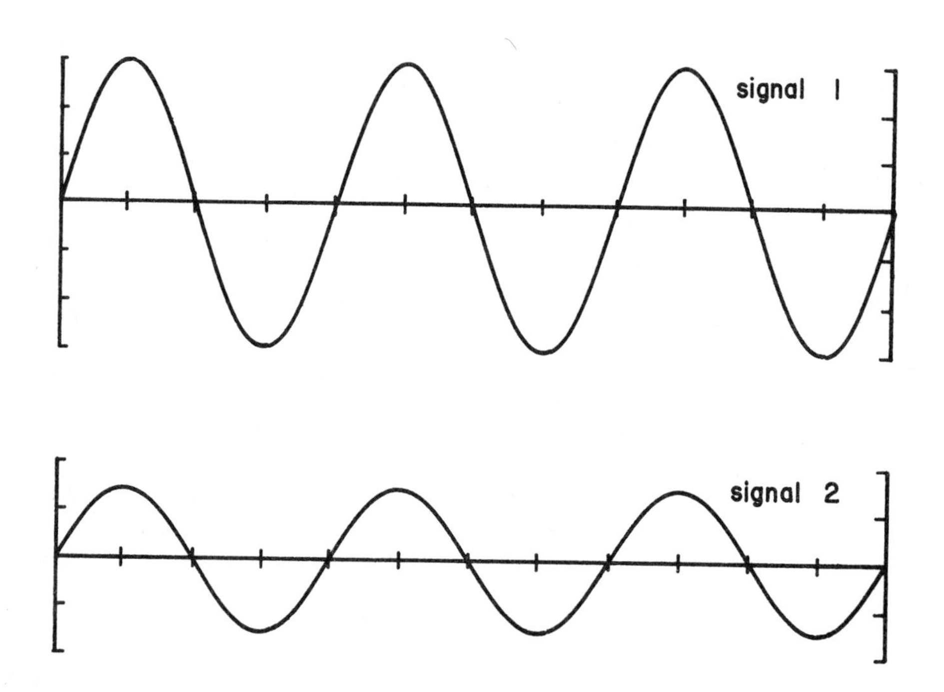

A sine or cosine curve can be sketched with relative ease when some obvious symmetries are used in the graph. One half-cycle can be divided along the horizontal axis into three roughly equal segments. The first and last one-third of each displacement is nearly a straight line. The central one-third is nearly a small radius semicircle. The straight line segments continue and connect each half-cycle. In most situations a sine or cosine graph can be replaced by a triangular graph of the same amplitude and repetition rate.

b. Equal frequencies with a different initial phase (reverse phase) than in part a and different amplitudes.

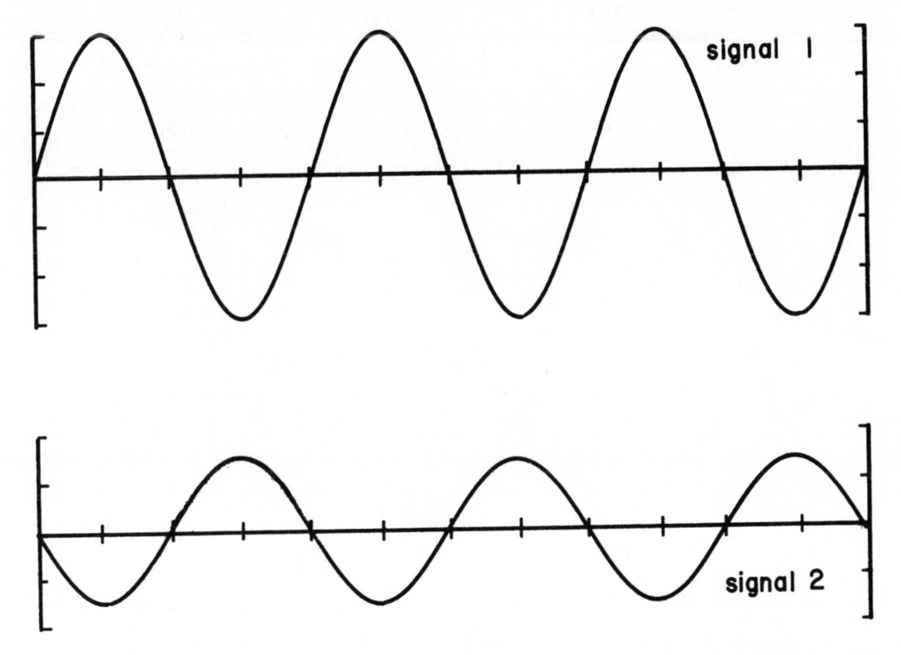

4-3. As a further illustration of the result of summing two vibrations find the composite graph for the two signals shown in the illustration. The two signals are of unequal amplitude, unequal frequency, and have reverse initial phase.

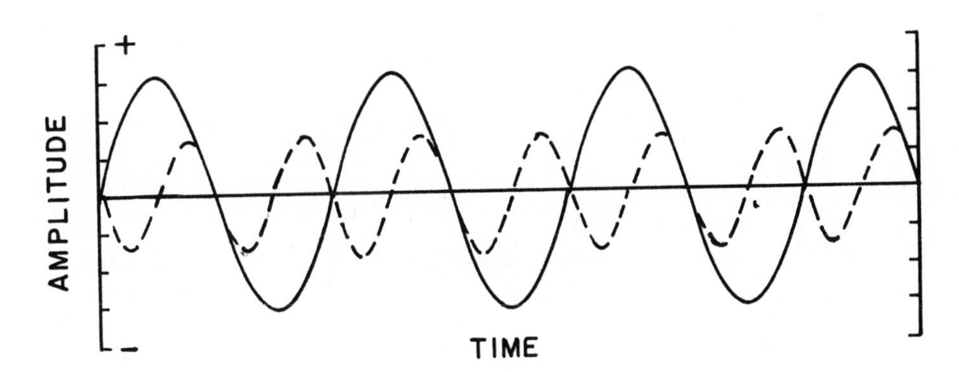

4-4. On graph paper sketch the curves that represent time variations in amplitude of two model oscillators. Carefully <u>add</u> the amplitudes of the two sinusoids. (The first has a frequency of 5 rep/sec, and the second has a frequency of 6 rep/sec.) Use a compass, dividers, or a ruler to transfer the displacements from equilibrium constructing the summation graph. Show several representative points. Is the resultant a sinusoid?

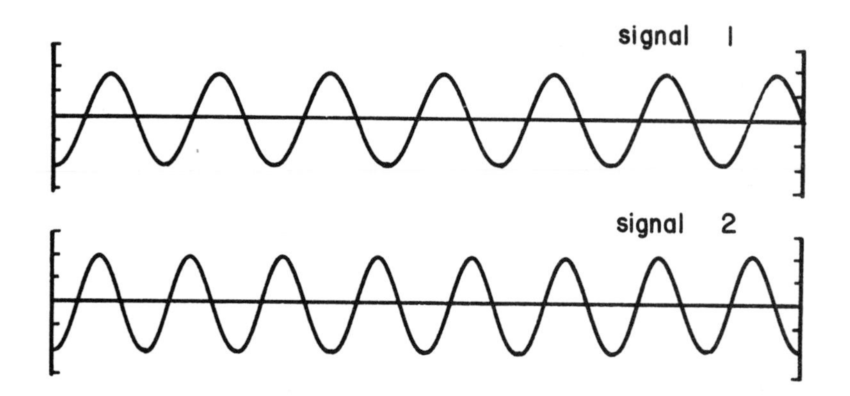

4-5. On graph paper sketch the curve that represents the summation of the time variations in amplitude of two model oscillators which individually oscillate as sinusoids. In contrast to the situation in problem 4-4, these oscillators have repetition rates not related by small integers. The duration of one oscillation of the slower signal is 1.414 times longer than that of the faster signal. What form would the summation graph have if it were continued for as many as fifty oscillations of the slower signal?

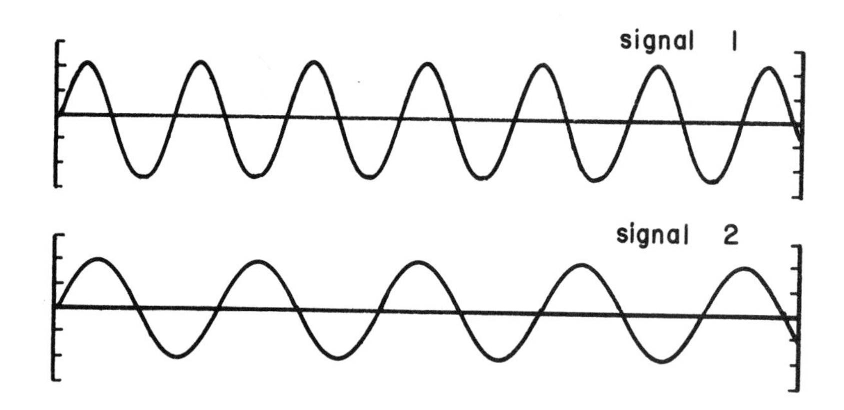

4-6. Consider the case where two sinusoids are not exactly in phase but are of equal amplitude and frequency. Write a brief description of what happens when one of the sinusoids progressively slides over from nearly the same initial phase to reverse phase.

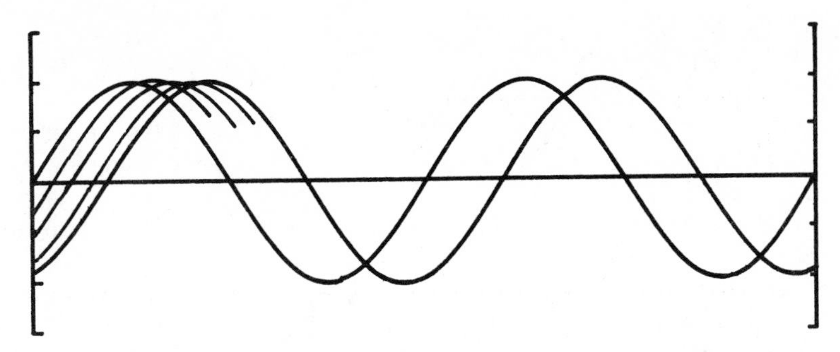

4-7. Not all oscillators have an amplitude which varies sinusoidally with time. Electronic oscillators can produce signals of the type graphed below. Sketch a curve of the summation of the signals. What is the repetition rate of the summation signal compared to the repetition rate of signal 1? Suppose that signal 1 were delayed by one-half the time between repetitions (moved to the right by one-half the time for one repetition). What is the repetition rate of the summation signal compared to signal 1? Something like this could occur in electronically-generated sounds when a "sawtooth" is combined with "square wave" signals electrically. (Particularly in this problem and in the others in this chapter, you are to resist attempts to discuss how the combinations of curves would sound if they were somehow made audible. If the combination signal were fed into a loudspeaker so that you heard the sound in a room, you would need to consider a number of interrelated phenomena that are a part of hearing, including the response characteristics of the loudspeakers, the excitation of room modes, the loudness of the sounds, and the heterodyning that occurs in the ear.)

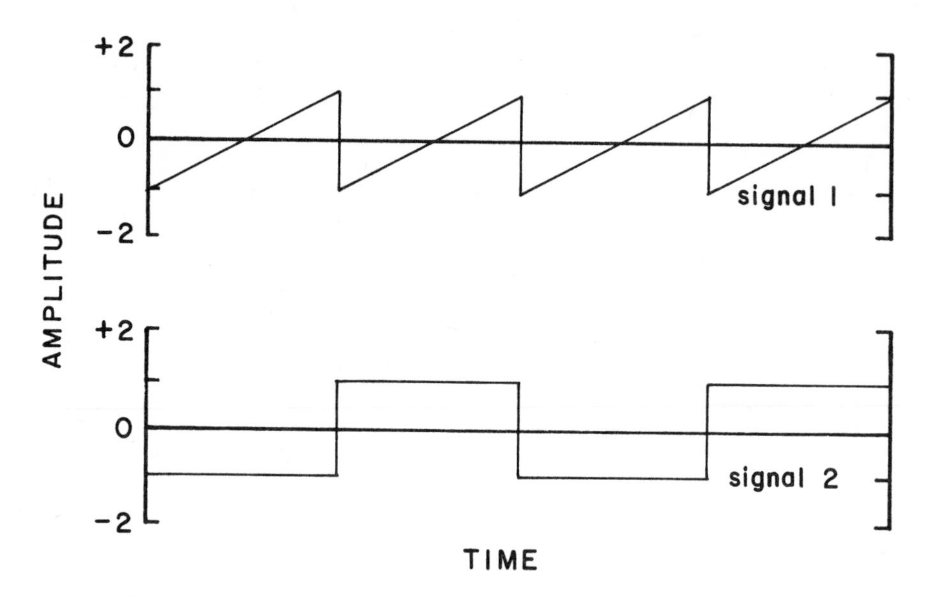

4-8. When one snaps his fingers the sound produced has a waveform very similar to that illustrated below. Find the repetition rate, the initial amplitude, and the halving time of the oscillation of a sinusoid of decreasing amplitude that could represent such a damped oscillation.

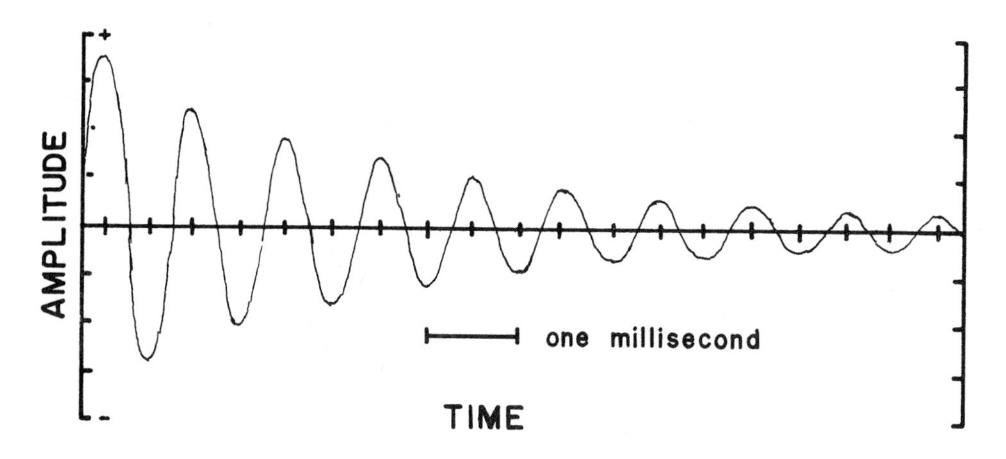

4-9. In order to further accustom your ears to complex sounds find an oscillator you can investigate by tapping on its various parts. Some suggestions are an empty coffee can, an empty roasted nut can, a metal water pitcher, a metal tray, or a cooking pan. Produce sounds by tapping the object in several spots and identify the pitch by ranking the sounds from the lowest to the highest, using a piano as a reference. Discuss the characteristics of the sounds by describing the repetition rates that seem to be present. Are they separately recognizable as single sinusoids, or do they each consist of several components? (At first the sounds may seem to be noise without a characteristic repetition rate. With practice you can sense individual repetition rates within these complex sounds.)

5

PITCH: THE SIMPLEST MUSICAL IMPLICATION OF CHARACTERISTIC OSCILLATIONS

5-1. The sounds generated in a touch-tone phone dialing system are combinations of the following:

697, 770, 850, 941, 1209, 1337, 1477 Hz

Each sound produced by touching a single button consists of the combination of two frequencies, obtained from Table 5.1 below. This table represents the placement of the buttons on the touch-tone pad and the frequencies associated with each row and each column.

Table 5.1

Frequencies Assigned to Buttons on a Touch-Tone Dialing Pad

Hz	1209	1337	1477
697	1	2	3
770	4	5	6
852	7	8	9
941	*	0	#

a. Use Figure 2.1 to find the nearest note names for each of these various component frequencies.

b. Which if any of these composite sounds are in fact tones in the restricted sense employed in <u>Fundamentals of</u>

 Musical Acoustics? Can you arrange any combination of these single frequencies to fit the classification as either exactly or approximately harmonic components?

5-2. When multicomponent sounds are received at the ear, under certain conditions the relative phase contributes very little to a change in pitch although the phase shift may contribute other qualities such as roughness to the sound. (This is especially true in a room. Why?) Use a pair of dividers or a scale to prepare a graph of the combination curve of the three curves given below.

a. For different frequencies with the same initial phase.

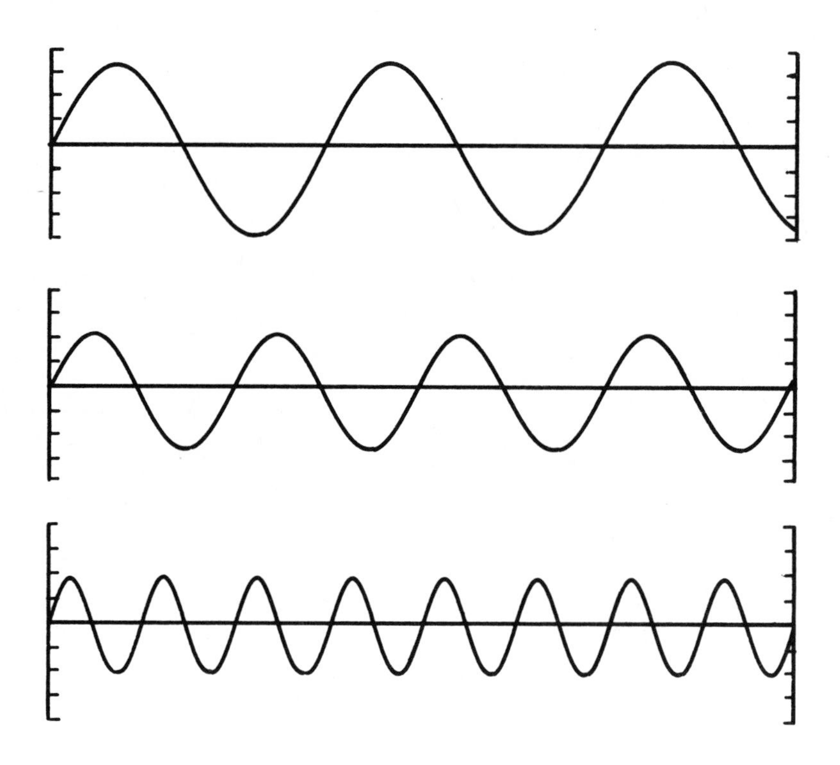

b. With different initial phases.

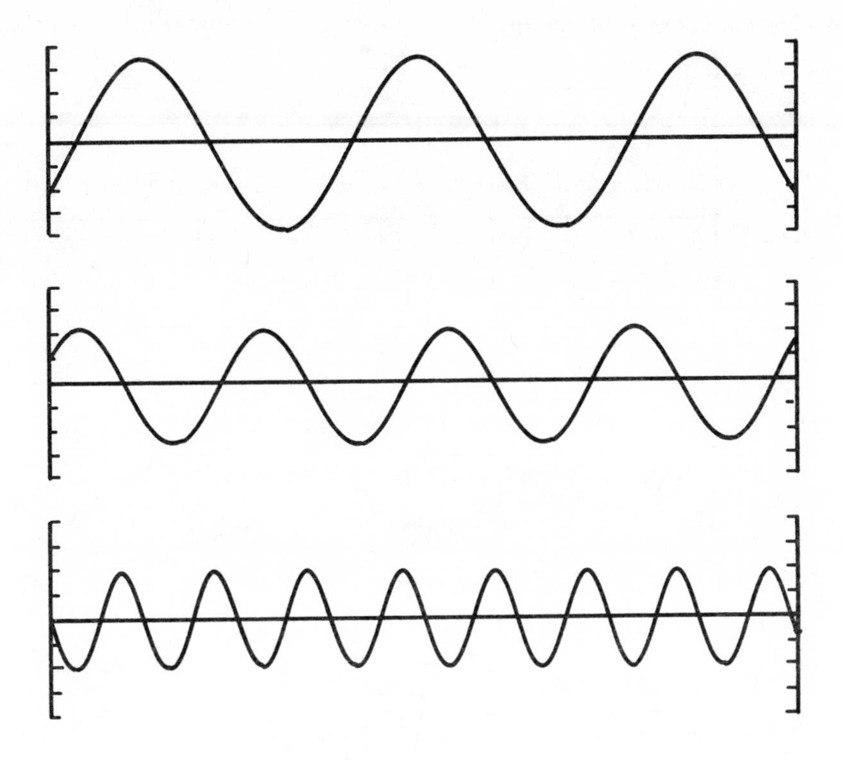

5-3. The following is a sketch of the oscilloscope trace of a sound.

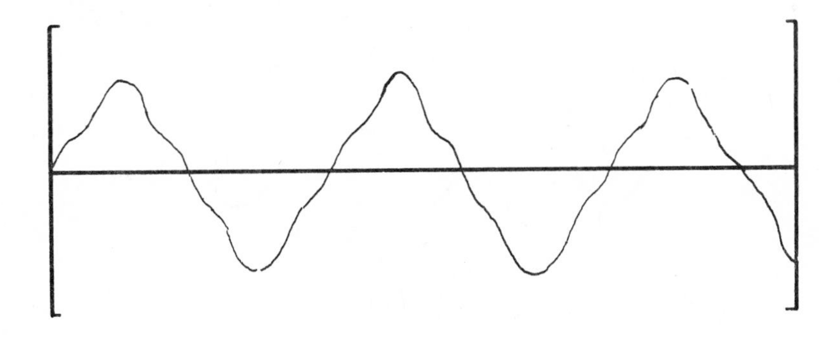

If the repetition rate (frequency) of the large amplitude is 500/sec, discover the rate of the small amplitude alteration superimposed on it. What requirements are made of the initial phase of the two component signals?

5-4. Now consider the following sketch of the oscilloscope trace of a sound.

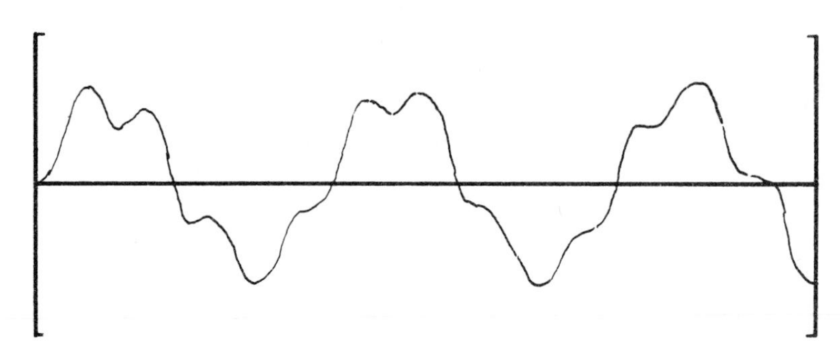

If the repetition rate (frequency) of the large amplitude alteration is 250 Hz, discover the rate of the small amplitude alteration superimposed on it.

5-5. The sound from a single source has a vibrational recipe that contains the following collection of reasonably strong components:

200, 305, 410, 480

a. What would be a reasonably good prediction for a pitch name that might be assigned to this sound?

b. What would be a suitable pitch name for the sound produced if the 480 Hz component is removed?

5-6. Consider a sound from a single source whose components have the following frequencies:

258, 390, 528

a. What would be a reasonably good prediction for a pitch name for this sound?

b. What would be a suitable pitch name when the middle component, 390 Hz, is removed?

5-7. The relative phase of the components of a sound can have an effect on the resulting wave shape. Make a sketch of the combined waveform which results from the summation of the four components given below. Use a ruler, compass, or dividers to transfer the amplitudes. As an aid in making your sketch superimpose the first two and then the next two before adding the intermediate curves to form the final superposition.

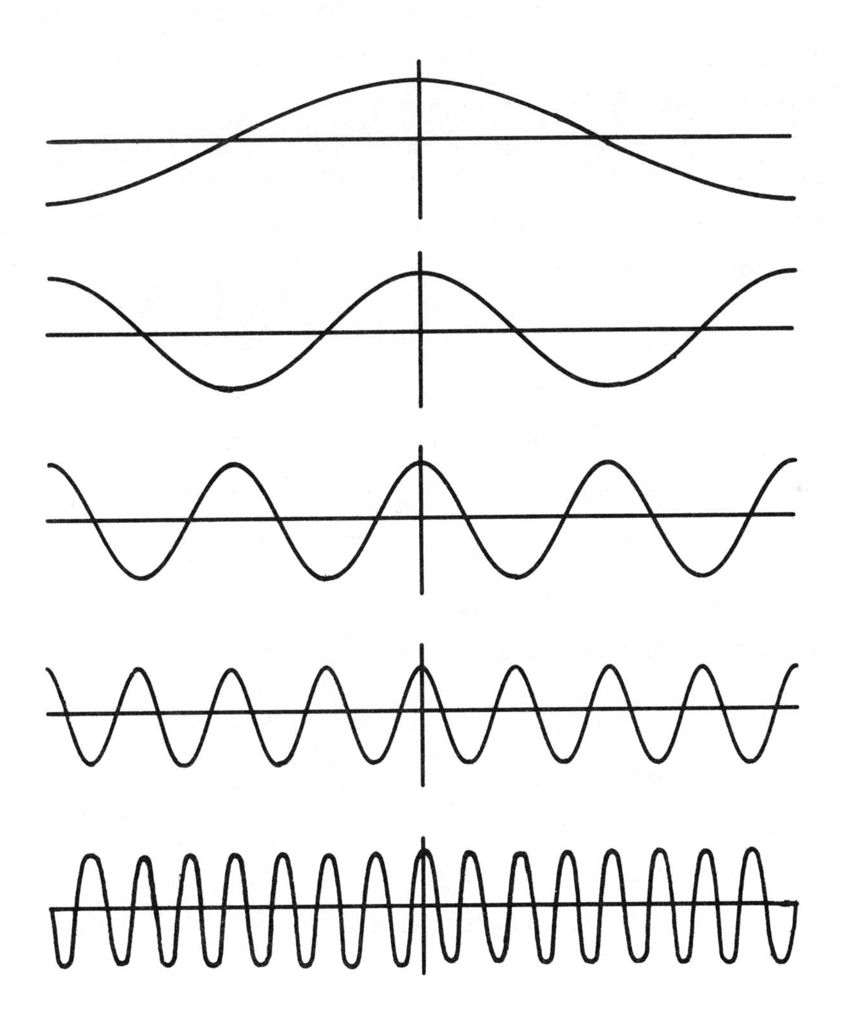

Now make a superposition curve of these component curves with the initial phases changed from the first case.

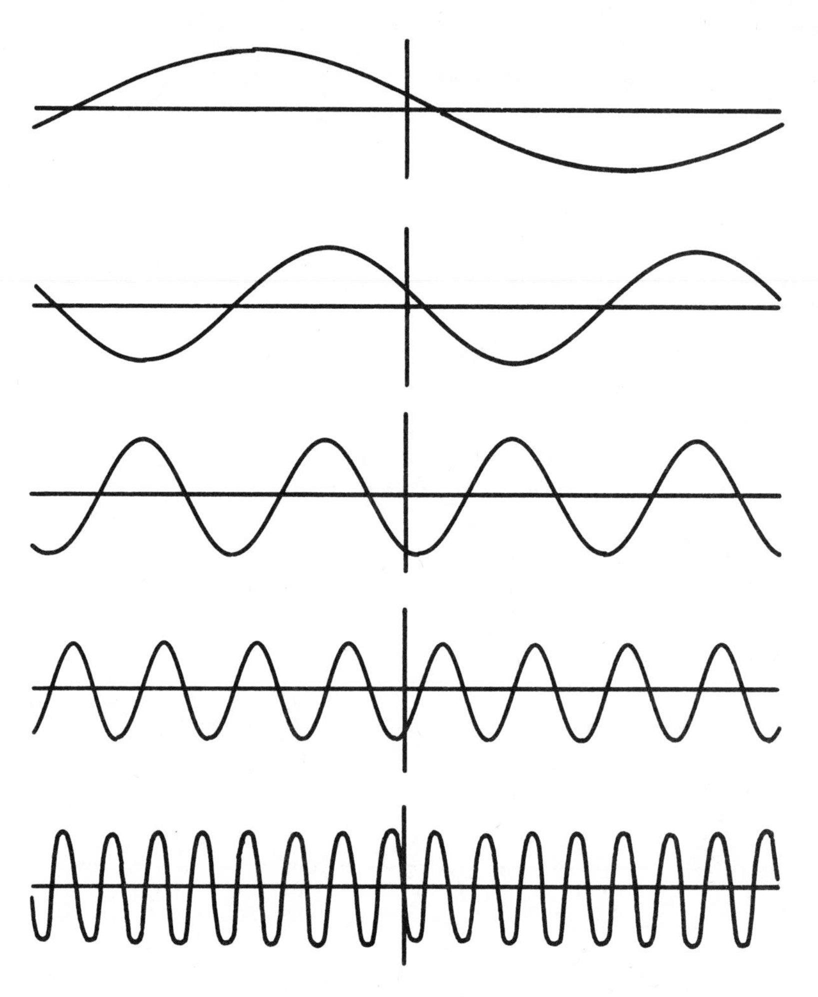

The whispered sound "tick" and "ssh" have the same general sort of amplitude and frequency components as these examples. For each case explain the difference in these sounds in terms of the frequency ratios and the phase relationships between the components.

5-8. A sound from a single source that resembles the twang of a banjo might have a set of frequency components such as

100, 199, 295, 385, 510, 605, 703 Hz

For a banjo-like sound the lowest few components (four in this example) die away more quickly (halving time is shorter) than do the upper ones.

a. Can you identify a relationship among these components that distinguishes them from the longer-lived ones?

b. Speculate on the reason why the pitch of such a sound is less well-defined than one put together from the components

100, 199, 295, 385, 490, 595, 697 Hz

The decay habits for this second sound are the same as that of the first example.

5-9. Consider a sound (single source) that has the following rather interesting recipe of the components

150, 170, 305, 340, 447, 510, 596, 680 Hz

Some people might hear this as a messy single-pitch sound (rather unlikely) or they might hear two pitches. Give pitch names for both cases. Explain the reasons for the selection of the pitch.

6

THE MODES OF OSCILLATION OF SIMPLE AND COMPOSITE SYSTEMS

6-1. Define the concept of uniform motion using examples of motion in a straight line and along a circle. Consider a bicycle or car moving at constant velocity. Note which part of the vehicle is in uniform linear motion and which part is in uniform circular motion. (Consider only the frame and the wheels.) Also describe the position an observer would have relative to the vehicle to see each type of motion.

6-2. It is a useful aid in visualizing single frequency sinusoidal motion to draw sine and cosine functions and relate them to the motion of a point that moves in a circular path. The motion can be divided into twelve equal time intervals. The vertical location of a point on the graph of the sine function is the height of a point on the circle above the horizontal line drawn through it. The vertical location of a point on the graph of the cosine function is the horizontal distance of a point on the circle measured from the vertical line drawn through it. Notice that the sine and the cosine are exactly alike except that one is moved over a quarter-cycle from the other. This observation can be the basis for constructing the cosine curve rather than transferring the horizontal distances used in the definition of the cosine. The sine function is zero at the initial step and grows in a positive direction away from the origin. The cosine function starts from unity and falls toward zero during the first quarter cycle. Notice also that each quadrant of a sinusoid is exactly the same shape.

a. Construct a graph of sine θ.

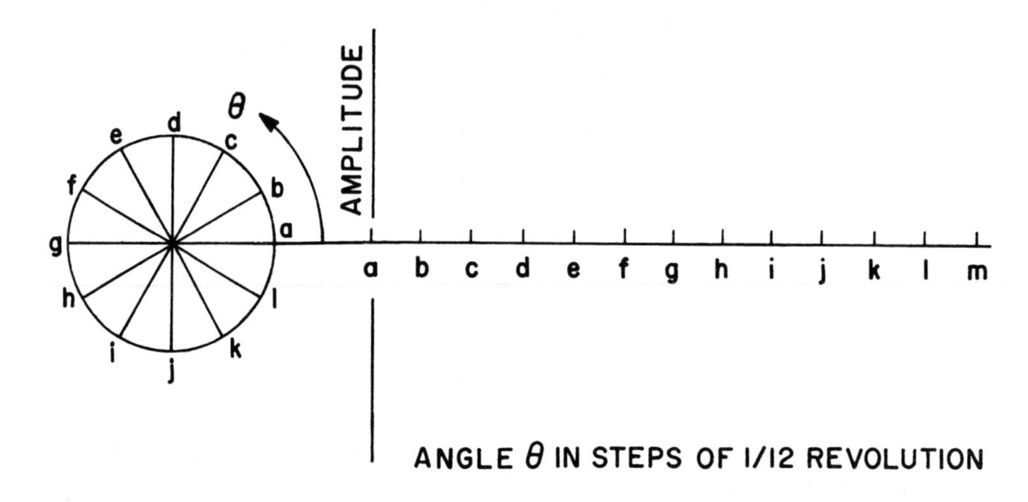

b. Construct a graph of cosine θ.

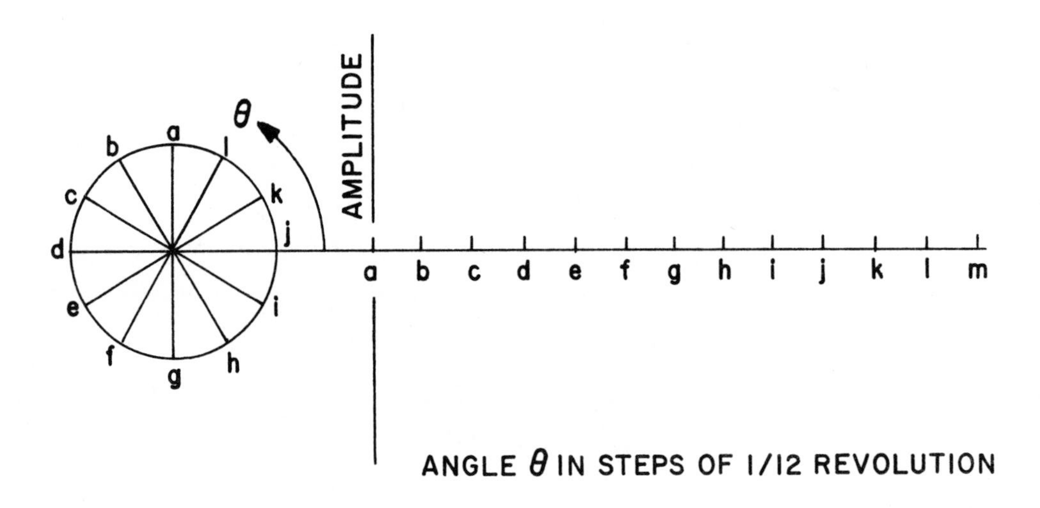

6-3. Consider the following oscillator systems. What is the ratio of the frequency of each oscillator compared to the one labeled a?

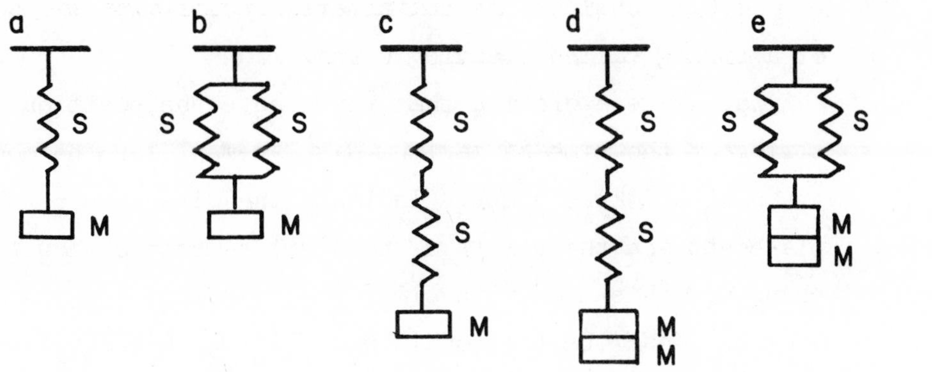

6-4. Describe a system oscillating in simple sinusoidal motion. Construct your own system using rubber bands for the elastic restoring force and some unbreakable object such as a large washer, metal or plastic ring, or plastic cup for the suspended mass. Set the object into vertical motion and measure the time for one oscillation. After a few trials with different masses you should have an oscillation that is slow enough for the amplitude to be measured by sighting onto a ruler held vertically behind the mass. Sketch the displacement for your system as a function of time. Use an estimate or measurement of the mass of the object along with the frequency of oscillation to calculate the stiffness of the suspension. When your system is set into oscillation, swinging from side to side as a pendulum, does it have the same period as you found for the vertical motion? Since the mass was the same for both motions, what is the stiffness for this second type of displacement?

6-5. Carefully sketch the <u>first five</u> modes of a tightly stretched string anchored rigidly at its two ends. (Be specific as to where the location of positions of no motion occur.) Regard the string as if it were a perfectly flexible string of uniform diameter and the anchors are so stiff that there is no motion of the ends of the string.

6-6. Consider the characteristics of vibrational modes of a string called standing waves.

a. Make a drawing that represents the position of the string at the time of maximum displacement for the mode whose graph shows three humps. Indicate the direction of the displacement and the points of no displacement to show the change of this mode with time.

b. Describe the conditions existing a quarter of a complete vibration later than the situation drawn in part a.

6-7. Consider the characteristics of vibrational modes in a narrow pipe open at one end and closed at the other that are called standing waves.

a. Make a drawing that represents the pressure along the pipe at the time of maximum pressure change for vibrational mode 2. Indicate the places along the pipe of pressure change (increase or decrease) and no pressure change in order to show where they change with time for this mode.

b. Describe the conditions existing a quarter of a complete vibration later than the situation drawn in part a.

6-8. What is the effect on the vibrational frequency of a string when the tension is changed by a factor of 2 in comparison with a change of the cross-sectional area by a factor of 2 (for the same material)?

6-9. A taut string is capable of vibrating in three different ways: transverse, longitudinal, and torsional. Describe the characteristics of each type of vibration and indicate ways in which they could occur in musical situations. Discuss the ways such properties of the string as mass, diameter, length, and tension influence the frequency of vibration.

6-10. Methods of graphical analysis can be used to verify the amplitudes given for the harmonic components of a square wave as obtained with a Fourier analysis. The vibrational recipe for the sinusoids to be combined is

Serial No. of Sinusoid	1	2	3	4	5	6
Amplitude Referred to the Fundamental	1	0	1/3	0	1/5	0

Construct a graph for the composite that results from the superposition of the sinusoids. Show one cycle of the fundamental and as many cycles of the high frequency members as are required to produce one cycle of the combined wave form. The sinusoids are all of the same initial phase.

6-11. Methods of graphical analysis can be used to verify the amplitudes given for the harmonic components of a sawtooth wave as obtained with a Fourier analysis. The vibrational recipe for the sinusoids to be combined is

Serial No. of Sinusoid	1	2	3	4	5	6
Amplitude Referred to the Fundamental	1	1/2	1/3	1/4	1/5	1/6

Construct a graph for the composite that results from the superposition of the sinusoids. Show one cycle of the fundamental and as many cycles of the high frequency members as are required to produce one cycle of the combined wave form. The sinusoids are all of the same initial phase.

7

Introduction to Vibration Recipes: The Plucked String

7-1. In order to gain additional experience in adding sinusoidal curves, consider a sinusoid with a definite amplitude and frequency, and add it to a second sinusoidal curve whose frequency is twice that of the first and whose amplitude is one-third that of the first. Let the sinusoids start out together with zero amplitude and for the first quarter-cycle have increasing amplitudes.

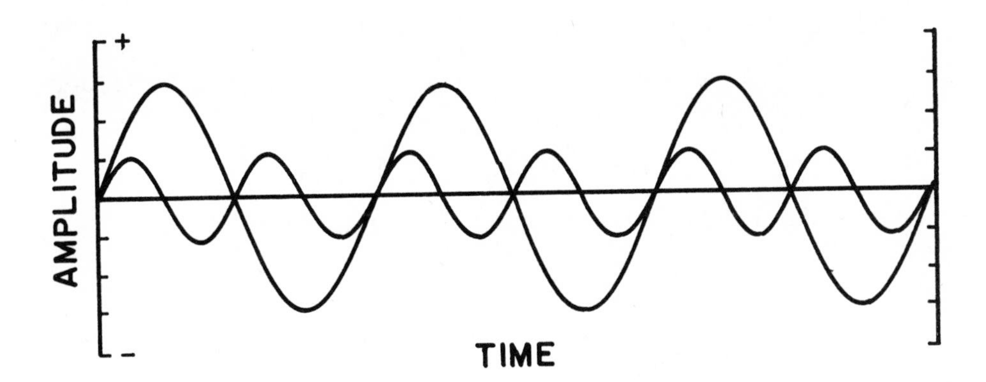

7-2. Develop the vibrational recipe for the situation graphed below. How much amplitude of mode 1 and how much amplitude of mode 2 are required to produce the following initial configuration of a two-ball spring-mass chain?

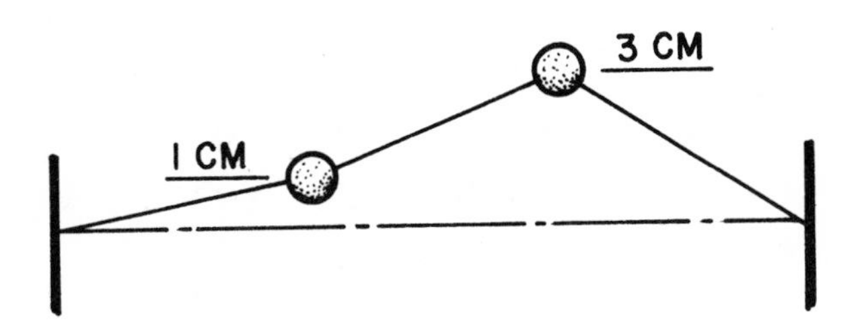

7-3. Develop the vibrational recipe of a three-ball spring-mass chain system for the initial configuration shown. First sketch a drawing of the configurations of the first three modes of a three-ball spring-mass chain system. Second determine how much amplitude of modes 1, 2, and 3 are required to produce the following initial configuration. (It is no accident that the configuration of mode 1 is suggestive of a sine function.)

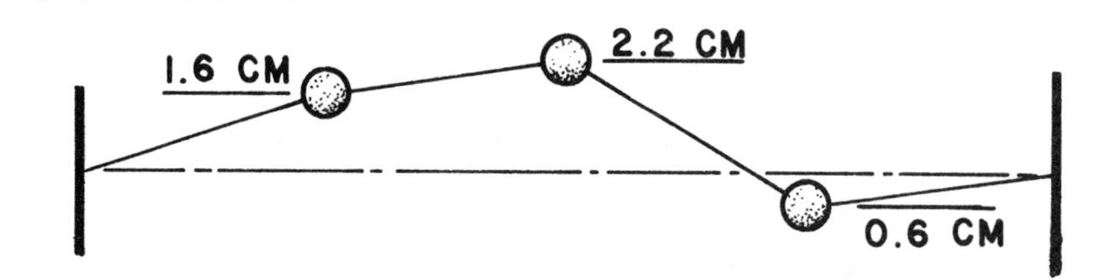

7-4. The vibrational recipes for struck and plucked strings involve the factors $1/n$ and $1/n^2$. Prepare a single graph of these quantities as a function of n for values of n ranging (as a minimum) from one to eight. A graph whose horizontal and vertical axes are 4 inches (10 cm) will show the more rapid decrease of $1/n^2$ compared to $1/n$.

7-5. What modes are absent when a string 32 cm long and fixed at both ends is excited at 4 cm from one end by a very narrow plectrum?

7-6. What happens to the vibrational recipe of a guitar G-string (open string pitch at G_3) that is plucked at any normal spot and then has a sponge or cloth damper applied momentarily to the midpoint of the string? If a similar experiment is carried out with the damper applied exactly one-third of the way along the string, what will be the difference in the vibrational recipe for the two situations? Suppose an additional experiment is done where the string is plucked one-fourth of the way along its length and the damper applied at this point. What would these changes in the vibrational recipe suggest about the perception of the sounds produced by the various manipulations of a string?

7-7. Consider a string that is 60 cm long and mounted rigidly at both ends. If the string is plucked with a broad plectrum of width W = 2 cm, what is the serial number of the mode for which no excitation is possible? Use the criteria that no mode excitation occurs when the width W equals one wavelength. Will the pluck point have any bearing on your answer?

7-8. Consider a harpsichord string that is plucked at a distance of exactly one-eighth of the way from one end with very narrow plectra. By means of mode diagrams, show that the 8th and 16th modes will not be excited at all. Show further that the 7th and 9th modes will be excited weakly. Show also that the 3rd mode is excited strongly for the plucking position described above. You can consider a harpsichord string 1 meter long and the pluck point to be deflected 4 mm before release if you wish to estimate relative amplitudes of the modes.

7-9. Since a real string has a certain amount of stiffness, its vibrational frequencies will differ somewhat from an "ideal" string assumed to have no stiffness. Discuss the nature of the expected changes in frequency. Relate the possible shifts in the partials relative to those of an "ideal" string. You are to consider the stiffness that the string has as a result of the resistance to bending and comment on the resulting vibrational recipe for the partials.

7-10. The term "wave" is used to describe oscillatory behavior observed for various materials. Descriptions of the different types of oscillations include the terms progressive (running) waves and stationary (standing) waves.

 Define these terms graphically and verbally with respect to both time and distance. Some writers assert that the term "standing wave" is a misnomer. Whether or not you agree that this is a correct viewpoint, discuss the definition of the term "wave" the writer seems to be using as the basis for his comment.

7-11. Suppose that a certain irregularly shaped string-like object is stretched between two fixed points so that it vibrates with characteristic modes of oscillation such as those shown in the sketches.

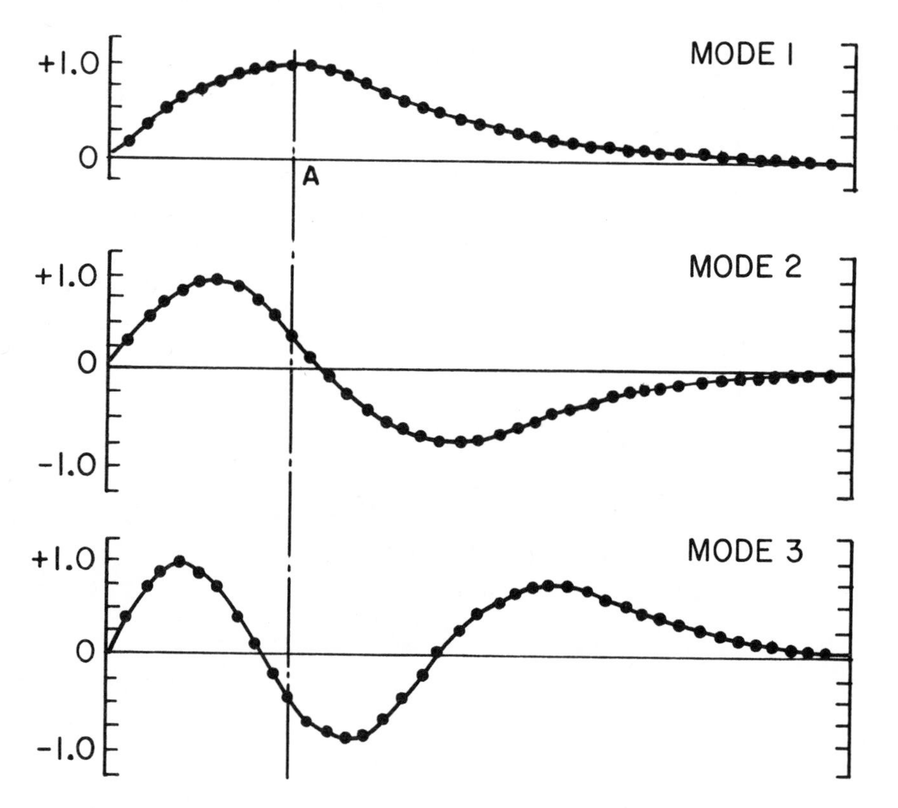

a. Calculate as accurately as possible the amplitudes of the first three modes of oscillation when the string is plucked at the point marked A.

b. Suppose a pickup is located one-fourth of the way from the left end of the string. What is the vibrational recipe for the displacement of the string at the pickup location when the string is plucked at point A?

c. What is the vibrational recipe for the displacement at the center of the string when a pickup is located there and the plucking point remains at A?

d. Then combine the two pickup recipes for the joint output of the two. Discuss the method you use to find the joint output. (This is similar to the way the output of two pickups are used together on an electric guitar.)

8

Broad Hammers and Plectra, Soft Hammers, and the Stiffness of Strings

8-1. Suppose that instead of striking or plucking a string at some point we touch it lightly at this same point in a special way. Assume all modes of a string are set into oscillation. Then we touch it lightly with a wisp of cotton, which we continue to hold in light contact, at a point one-sixth of the way from one end. With what relative quickness will modes 1, 2, 3, and 6 decay away? You may use the adjectives "quick," "medium," "slow," etc. in answering this question. Try to attach numerical values to these damping-rates in terms of the relative halving times (the time for the amplitude to decrease by one-half).

8-2. As a contrast to the vibrational recipes for a plucked or struck string consider the effect of attaching a tiny blob of wax to the string. Assume that all of the modes can be set into motion. Use your general knowledge of the factors that determine the vibrational frequency of anything to say whether such an addition of a small blob at one-sixth the distance from one end will raise or lower the frequencies of the various modes (1, 2, 3, and 6 in particular). Attach adjectives to the various mode numbers telling the degree to which you expect the individual frequencies to be altered.

8-3. A pencil and paper experiment can be conducted to illustrate the relationship between the impact duration of a hammer and the vibrational recipe. Consider the first ten modes for a struck string where the amplitude decreases with a reduction factor of 1/n for increasing mode number. (Refer to problem 7-4.) Suppose that the frequency of mode 1 is 250 Hz and the higher frequencies are harmonics of 250 Hz. Which modes are excited almost exactly as if the rebound were instantaneous; which modes are excited about half as strongly; and which modes are excited almost not at all for each of the three impact durations of 0.0002 sec, 0.0004 sec, and 0.001 sec? Suppose the characteristic time for contact was 0.04 sec; which modes would be excited?

8-4. Suppose that a string 45 cm long is plucked with a rather broad plectrum 1 cm wide. Find the mode for which there is no excitation. If one considers the 1/n factor (refer to problem 7-4), the modes for which the excitation is very weak lie in what range of serial numbers? The effect of the pluck point on the vibrational recipe can be combined with the cutoff in an estimate of the relative strength of the modes of a plucked string. Draw either a single graph or three graphs that show the effect of plucking this string with the wide plectrum at the center, one-fourth the way from one end, and one-tenth the way from one end.

8-5. Choose some topic having to do with plucked or struck strings, such as width of plectrum, width of hammer, striking point, hardness of plectrum, and write a paragraph on some of its musical implications. Do not attempt to deal with all of these aspects—just one. You may find it interesting to contrast two instruments, or you might prefer to deal with differing aspects of only one. For example, the plucking position or type of plectrum used to play a banjo might be an interesting topic.

8-6. The actual dynamics of piano hammers are somewhat complex. Under certain conditions the string throws off the hammer more quickly than one would expect from a study of hammers of this type striking a solid object. Because of the graded nature of hammer-felt and other subtleties, one also finds that a heavy blow tends to cause the hammer to rebound in slightly less time than is the case when the blow is softer. (Note: This is all contrary to the predictions of straightforward elasticity and standard wave theory.) One can deal with these phenomena mathematically, and for present purposes there is no difficulty at all in predicting what changes in spectrum are to be associated with the shortened rebound time. Make a brief prediction of the vibrational recipe and give a short summary of your reasons.

8-7. One of the ways in which the selective damping technique can be used to check the effects of the plucking point on the excitation of string modes is to consider some interesting locations along the string. Consider a string set into vibration initially with all modes of oscillation. The string is now touched with a damper at points such as $1/\sqrt{2}$, 9/100, or $1/\pi$ of the length of the string. What would you expect the damping of the various modes to be for this situation? Would any modes die quickly to zero? Give reasons for your answers.

8-8. The perceived pitch of a vibrating string (and hence the fundamental frequency of the harmonic partials it is emitting) may be changed or selected by methods other than changing tension or composition. What is the relationship between the modal vibrational recipe for a string touched lightly at some integer fraction of the way from one end (for example, 1/2, 1/3, 1/4, etc) and the modal vibrational recipe for either part of a string if a narrow fret or bridge were located at the touch point?

9

The Vibrations of Drumheads and Soundboards

9-1. Sketch the first four modes of vibration of a circular disc having flexible or hinged boundaries with no stiffness at the edge. Indicate the general shape of the humps and show whether they are up or down with a plus or a minus sign. Show the locations of little or no motion.

9-2. What is the maximum possible frequency reduction of mode 1 of a circular disc held at the edge so it is hinged with no stiffness when a lump of wax is added to the center? Consider the mass of the plate to be 100 grams and the wax to be 6 grams. If the vibrational frequency of mode 1 is 440 Hz before the wax is added, what is its new frequency?

9-3. Why will the frequency of vibration of a tuning fork rise or fall when a piece of sticky material such as gum or clay is attached to the end of one of the tines?

9-4. Suppose one wished to adjust the frequency of vibration of a tuning fork and decided to grind off a small amount from the free end of each tine. Will the frequency rise or fall? Why?

9-5. Suppose the experimenter, in making the adjustment described in problem 9-4, removed too much material. What could be done to the tines to restore the previous frequency of vibration? Aside from the obvious addition of mass to the ends of the tines, what other change could you suggest to bring the fork back to the original frequency of vibration?

9-6. Beats are often demonstrated by detuning one of a set of matched tuning forks. How is the temporary detuning accomplished and what principle of vibrating systems is used in the process?

9-7. In addition to developing impressions of the pitch of sounds obtained by tapping on bars, plates, bells, guitar tops, etc., such tap-sounds can be used to give information about the duration of sounds. Arrange to observe some tap-sounds from bell-like objects. What property of the bell is measured by the time it takes the amplitude of the vibrations to decay by one-half? How would you go about getting this time if only your ears and a stopwatch were available for the measurement? Arrange an experiment so that you can estimate the halving time for the amplitude of vibration of a bell-like object. How many cycles of vibration are required for your source to reach the halving time? Occasionally this time is dependent on the initial amplitude. How would you check for this possibility? A piano or other musical instrument can be used to estimate the frequency needed for the calculation of the repetitions occurring in the halving time.

9-8. The top plate of a banjo may be regarded in a first approximation as a drumhead. In this approximation the bridge may be considered a mass that loads the drumhead at the place where it rests.

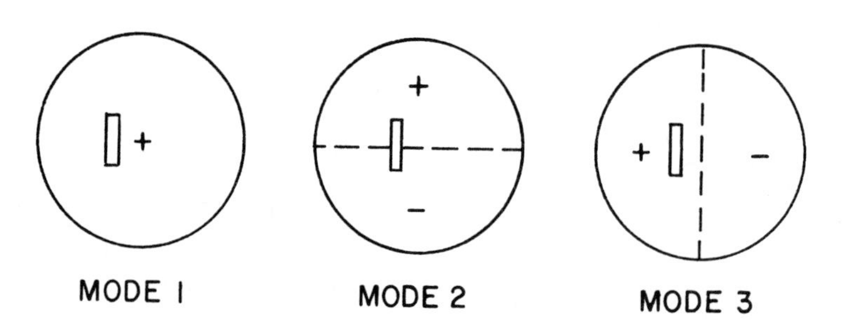

a. First, ignore the effect of string tension and sketch your idea of the humps for the first three vibrational modes of the bridge-plus-head system.

b. Second, describe the effect of string tension on the vibrational modes discussed in part a.

c. Next, devise experiments for tapping and selective damping to measure these natural frequencies. If you can arrange to observe these frequencies with the help, perhaps, of a piano, include a report of them.

9-9. For a mathematical look at vibrations of a rectangular sheet one can consider the modes f_{mn} that have their origins in the frequencies f_m and f_n via the equation

$$f_{mn} = \sqrt{f_m^2 + f_n^2}$$

where m and n are integers. Suppose that the frequencies are members of a harmonic series

$$f_m = mf_a \qquad \text{and} \qquad f_n = nf_b$$

which is not an entirely unreasonable circumstance. Continue the supposition by considering a "square situation" with $f_a = f_b = f = 100$ Hz. The algebraists will recognize the rewritten formula in the form

$$f_{mn} = f\sqrt{m^2 + n^2} = 100\sqrt{m^2 + n^2} \text{ Hz}$$

Locate a calculator equipped with a square root key and crank out the set of frequencies from $f_{10,10}$ to $f_{15,15}$. (Note: Because of symmetry you don't need to do them for all combinations.) Now from the list determine the difference in the frequency peaks; you can discover if the spacing is uniform or nearly so. What is the average spacing of peaks for this range of m and n?

9-10. Nylon is a material that "work hardens." As the material is stretched the internal structure changes and its strength increases. The property gives noticeable improvement in strings made from nylon. When a slightly non-uniform guitar string is pulled up to tension the thinner or otherwise weaker parts of it stretch, altering the internal structure of each portion and thereby strengthening it significantly. As the string is tightened further the relatively thicker or stronger portions are now stretched until they too are sufficiently work hardened. This effect can be used to stabilize the string. Each newly installed string is pulled individually about one and one-half semitones sharp for a few minutes before slackening it back to pitch. A noticeable improvement in the stability of tuning results, welcomed by guitarists.

a. Not only is the tuning made more stable by the prestretching described above, but the tone of a plucked string treated this way is perceived generally to be somewhat clearer. Explain why this should be the case with an analysis that is based on your knowledge of pitch perception and the effect of non-uniformity on the characteristic vibration frequencies of a non-uniform string.

b. A somewhat more subtle question arises from the observation that prestretched strings tend to produce a more sustained tone, as heard by the listener. Speculate on possible reasons for this, while keeping in mind the fact that the perceived audibility time and the physical halving time, although related to each other, are not simply proportional to one another.

9-11. A calfskin head for a kettledrum is a rather expensive object so a kettledrum may be fitted with a plastic skin. Some of them may have material sprayed onto a portion of the plastic sheet. The additional material is in the form of a ring about 2 inches wide located near the outer edge of the drumhead.

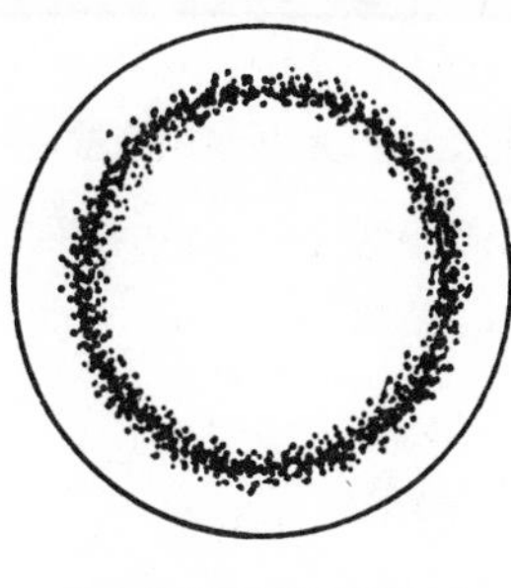

TREATMENT A

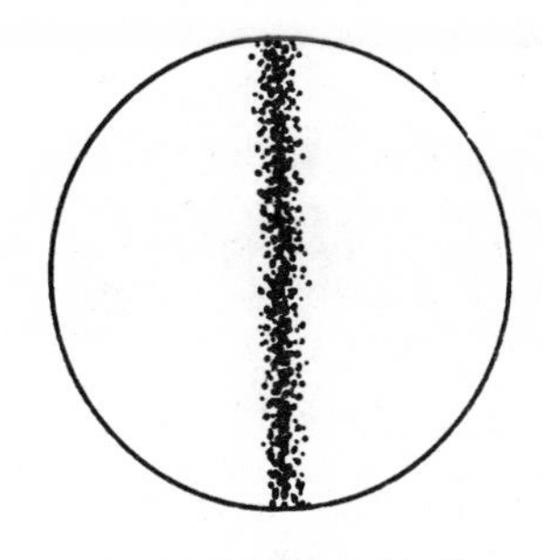

TREATMENT B

a. What effect might this have on the modes of the kettledrum?

b. Speculate on the possible effects of spraying a strip about 2 inches wide along a diameter of the plastic skin. Include any possible effects on mode frequencies and strike-point.

9-12. Refer to problem 5-8, which is a consideration of the pitch of banjo-like sounds, and problem 9-8, which is a discussion of the top plate of a banjo. Outline your present ideas of the physical causes for the frequency of plucked string components and the damping behavior presented in problem 5-8. Consult Chapter 7 of Fundamentals of Musical Acoustics by A. H. Benade or problems 7-7, 7-8, and 7-9.

10

Sinusoidally Driven Oscillations

10-1. Sketched below are the first four modes of vibration of a guitar top plate. The plate is installed on a guitar body so it is fastened at the edges. It will be helpful to read all parts of this question before working on the individual parts.

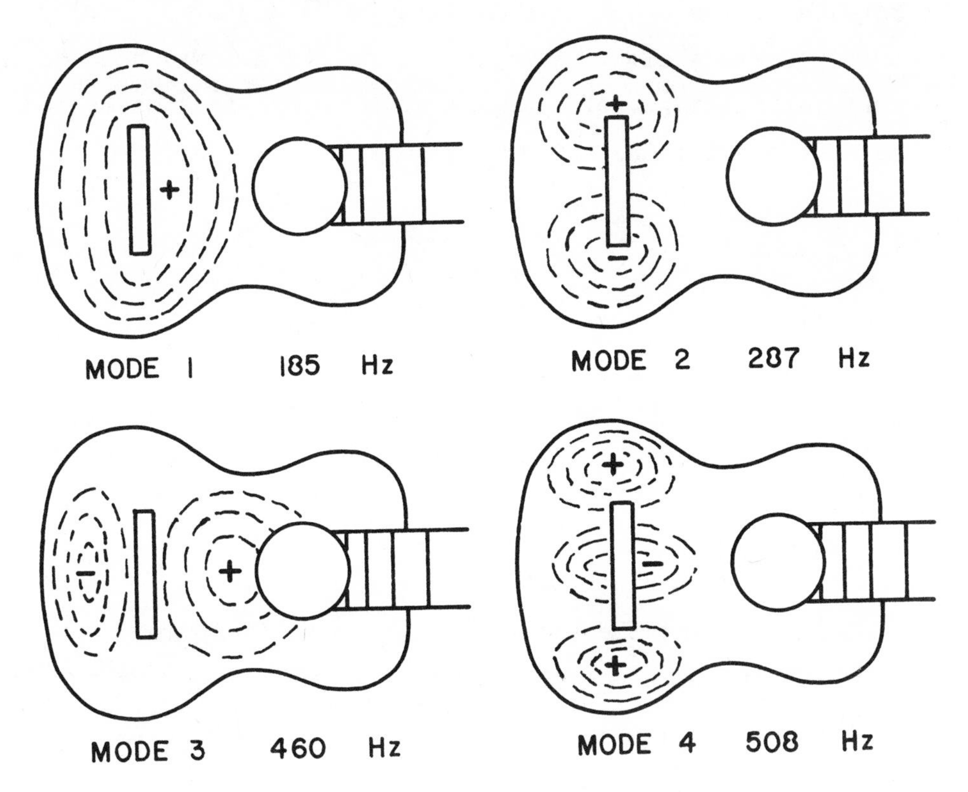

(Adapted from Fundamentals of Musical Acoustics, Arthur H. Benade, Oxford University Press, 1976 with permission.)

a. The bridge undergoes four types of motion: (1) rotation about the long axis of the bridge (see part b), (2) movement up and down perpendicular to the soundboard,

(3) rotation about an axis directed along the string so that one end of the bridge moves up while the other end of the bridge moves down, and (4) movement as a single unit parallel to the plane of the guitar top plate. These motions can result from the driving force imparted to the bridge by the string. Correlate these types of bridge motion with the aspect of string motion that drives them. A diagram is helpful but should not be used as a substitute for a brief description.

b. One of the ways for the bridge to move is in a rocking motion about an axis perpendicular to the strings, as diagrammed below. Discuss the relative influences of the four various types of bridge motion on the four modes indicated above for the guitar top plate.

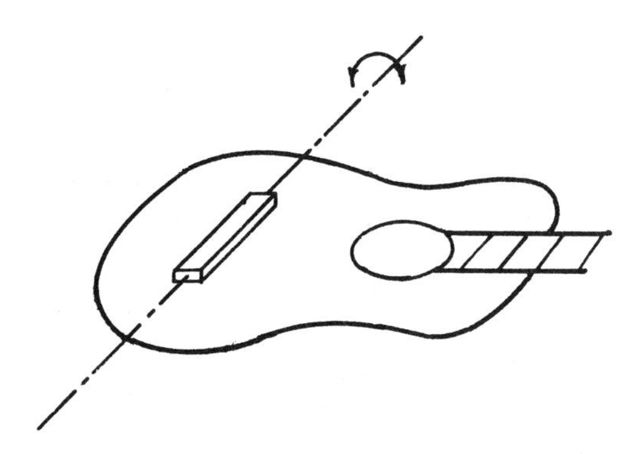

c. Mode 4 of the guitar top plate involves an additional motion of the bridge. Discuss how this mode can arise from the influence of the string. Which named note should be played on which string in order to favor this top plate mode?

10-2. Pretend that the guitar has a tall bridge of the sort used on violins and decide which named note played on it will drive mode 2 of the plate.

10-3. The air inside the guitar has its own natural frequency (the first mode of the air space) that lies near 90 to 100 Hz. In this mode the air squeezes in and out of the sound hole reminiscent of an oil can or plastic white-glue dispenser. Which of the plate modes would serve to drive this air oscillation with reasonable effect? (It is to be understood that the wide difference between the plate mode and the air oscillation frequencies will result in a relatively small mutual influence.)

10-4. Consider the situation where one experimented with a vibration pickup to detect the frequencies of the plate of the guitar discussed in problem 10-1. Suppose that the pickup was located at the position marked A in the diagram below.

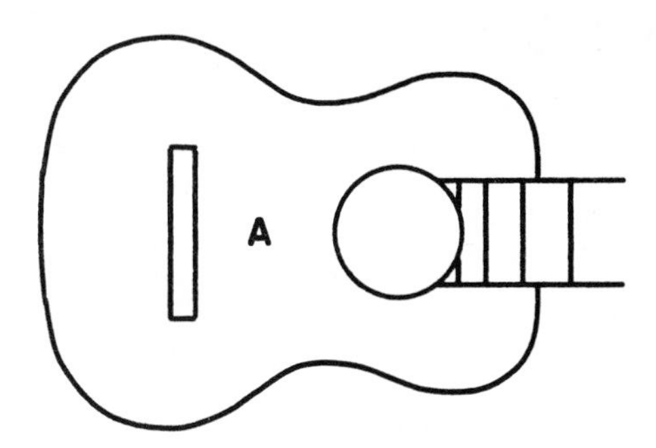

Assume the effective mass of the plate with the bridge attached is 325 grams. The vibration pickup adds a moving mass of 15 grams to the plate. Which of the modes will most probably be influenced? Estimate the shift in frequency for them. One manufacturer can supply a miniature vibration pickup that has a mass of 0.3 grams. When this pickup is attached to the guitar plate what shift in the frequency would be expected? How big are these shifts when referenced to the appropriate named notes?

10-5. Look at the vibrational shapes sketched in the diagrams in problem 10-1. Suppose you wished to measure the frequency of mode 2 by comparing its sound with that of some piano notes. Where would the plate be damped by touching with a finger, corner of a sponge, or wad of kleenex to damp the other vibrations? Where would you tap the plate so that mode 2 would be excited? You can use Figure 2-1 to find the named note of the corresponding vibrational frequency. This method can be used for an actual guitar but the frequency you find for this mode may be different. Guitars are constructed in many different ways so that the frequencies for all the modes will probably be different when compared to those given in problem 10-1.

10-6. Consider a vibrator of the spring-and-mass type with the characteristic resonance curve shown below. First draw a similar resonance curve. Then for comparison draw the expected resonance curve of an oscillator of the same stiffness (spring constant) and twice the mass. Also draw a curve of an oscillator of the same mass as the original, and twice the stiffness. Draw your curves as if the force of excitation were the same in all three cases.

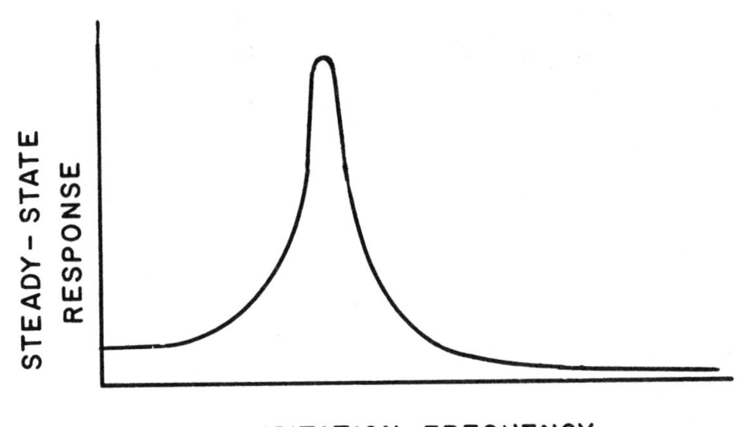

10-7. A simple oscillator of the spring-and-mass type constitutes a system capable of vibrating at a single frequency. Suppose both the mass and stiffness are increased by a factor of two. What will be the change in resonant frequency?

10-8. Explain how you could utilize the phenomenon of resonance for the purpose of determining the frequency of a tuning fork.

10-9. Distinguish carefully between "broad" and "sharp" resonance in terms of the ratio of the amplitude of the response to the bandwidth so that your definition does not depend on the magnitude (or strength) of the response. List examples of each type from your own experience.

10-10. List situations where resonators are used to improve or limit your ability to hear everyday sounds? List some practical purpose of resonators in your acoustical environment.

10-11. Construct a simple pendulum from a piece of light cord or thread and a mass such as a ring, washer, or similar object. An appropriate length for the string is 12 inches or 30 cm.

a. Suspend the pendulum so it is just above a flat surface upon which you can place a piece of paper. Mark the "at rest" position on the paper. Displace the pendulum bob about 2 inches or 5 cm from the equilibrium position and release it so it swings freely without external influences over the paper. Mark the position of the bob for the initial amplitude and the amplitude for each swing to the initial side of equilibrium. Time the swings of the bob to determine its natural frequency. (Hint: The time for one cycle is best obtained by counting for 10 repetitions.)

b. Construct a graph of the motion of the damped sinusoidal vibration of the oscillator in which the axes are labeled in agreement with your measurements.

c. From the data and the graph determine the halving time of your oscillator.

d. The damping of an oscillator and the decrease in amplitude with time can be used to obtain a measure of the steady-state response of a driven oscillator. From your data on the decay of the oscillations find the number of oscillations required for the amplitude to become one-half of its initial value. Use the value for the number of vibrations N to find the percentage bandwidth of the oscillator with the aid of the relationships

$$\text{PBW} = 38.2/N \text{ (percent)} \quad ,$$

and

$$W_{1/2} = 3.8/T_{rev} \quad .$$

For this study you can also find the frequency range of the resonance peak at half-amplitude (the half-amplitude bandwidth $W_{1/2}$). Decide if your oscillator has small, medium, or large damping.

e. Suggest a way to increase the damping of your oscillator that does not appreciably alter the period. Test your idea by altering the oscillator. For example, lengthening the support thread "slows down" the oscillator but does it affect the damping? What does lengthening the string change?

10-12. a. A particular piano is reported to have an F_3 string for which the halving time is 1.2 sec. Compute the number of vibrations (repetitions) of the string in this period of time. With the use of the relationship

$$\text{PBW} = 38.2/N \text{ (percent)}$$

find the percentage bandwidth for this string and the corresponding frequency range.

b. As a first approximation, one can assume that lower modes of the F_3 string and the neighboring G_3 string have about the same halving time. Explain why the G_3 string and not the F_3 string can be set into resonance by striking the G_4 when the felt dampers are not resting on F_3 and G_3.

10-13. In Fundamentals of Musical Acoustics by A. H. Benade, section 10.7 discusses the transfer response of a serving tray. Consider that material and the location of the resonance peaks in Figure 10.14 of the same book. The tray behaves in vibration very nearly as if it were a plate clamped at the edges. For practice in understanding this type of vibration you can conduct an analogous thought experiment.

a. Discover a general property of the peaks by estimating the average spacing of the characteristic frequencies.

b. Make one or two sketches analogous to the curves in Figure 10.14 for a rectangular membrane. The size and tension of the membrane are selected to give the same average spacing for those characteristic frequencies near 100 Hz as those found for the metal tray. As an aid in visualizing the experiment, sketch the membrane and indicate the locations assumed for the driver and motion sensor.

10-14. Consider the following hypothetical guitar-like instrument built solely for this example. The methods and procedures will have application to a real guitar. The top plate has a width that is half the length. Some representative modal patterns for the top plate are shown below.

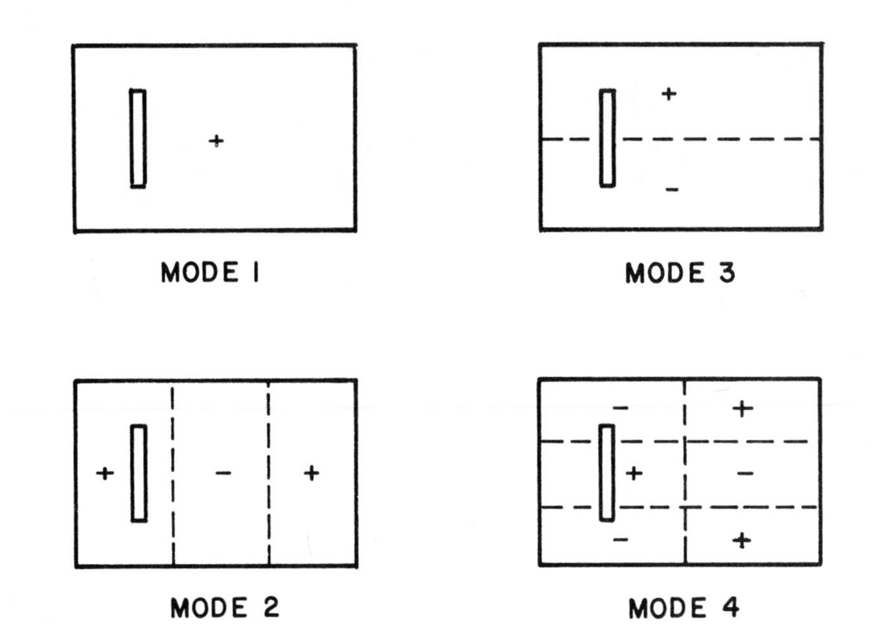

The vertical excitation of the bridge by the vibrating string is the means designated type A for the transfer of energy. The vibrational energy is transferred to a rotation about the long axis to the bridge by a means designated type B. Another possible means for energy transfer from the string to the soundboard is a rotation of the bridge about an axis directed along the long axis of the plate designated type C.

a. Discuss the relative amount of each type of excitation needed for the four top plate vibrational modes.

b. Refer to problem 10-1 for a discussion of bridge motion and the modes associated with each characteristic motion. Make an analogous discussion of the coupling of string motion, bridge motion, and top plate modes for this hypothetical instrument.

c. Assume that the frequency of mode 1 is 125 Hz. Also assume that all the modes are of higher frequency than mode 1 by the factors of 1.61 for mode 2, 1.84 for mode 3, and 2.83 for mode 4, where all the frequencies are for the plate without the bridge. The mass of the plate is 250 grams and the mass of the bridge is 20 grams. Estimate the resulting shifts in frequency of the various plate modes when the bridge is attached to it.

10-15. Construct a compound pendulum by using two identical objects such as steel nuts, metal rings, or washers, and lightweight but strong thread or string. Fasten one mass to a support with a piece of string of about 20 cm length. Now attach a second mass to the first with a string of the same length. Set the system into small amplitude oscillation in each of its modes and determine the frequency of oscillation of each such mode. The force should be applied to the second mass with a long, thin rubber band or elastic thread.

11

Room Acoustics I: Excitation of the Modes and the Transmission of Impulses

11-1. In previous problems you explored the nature of sounds initiated by tapping on an object and characterized the sounds with descriptive terms. Conduct a similar experiment in a medium-sized room by exciting the air into its oscillations with an impulse sound. A classroom, lecture hall, hallway, or recreation room are suitable choices. For the impulse sound you could burst a balloon or clap your hands at several locations in the room. Report any special "ringing," echoes, or long decay times, along with other descriptors that seem appropriate for the characterization of the sound.

11-2. You can test the general principles of the interchangeability of source and receiver by observing sounds in a medium-sized room. For the sound source a small radio, tape recorder, or signal generator with speaker is appropriate. Report the results of interchanging the locations of source and receiver. Describe the similarities and differences in the sound for interchanges that involve several pairs of locations in the same room. You may obtain interesting results from studies of several rooms. (Caution: Should you find noticeable differences for exchanges of the source and you as receiver in certain rooms, do not conclude in haste that the interchangeability principle is incorrect. Since you had to move to the new location along with your ear, the room modes were altered somewhat.) The experiment can be repeated with a tape recorder and single sound source so as to illustrate the effect. Listening to the recorded signals would give a better impression of the effects of interchanging source and observer. You could remain in the same place in the room for each recording.

11-3. It is useful to make a reference list of characteristic distances for the time of travel of sound. Assume that the speed of sound is 345 meters/sec and compute the distance a sound travels in 1 millisec, 5 millisec, 45 millisec, 0.1 sec, 1 sec, and 5 sec. As an aid in remembering these distances and times, list some approximate distances such as the length of your foot, your height, a car length, or city block that can give easy-to-remember values of either time or distance.

11-4. The speed of sound is the same for all frequencies; therefore all components of a distant sound arrive at the same time. In an open field where there are no reflections or room modes to complicate things, what would a person expect to hear when a major triad were played some distance away from the observer if the speed of sound were proportional to the frequency? The distance is imagined to be great enough that a half second or longer is required for the sound to reach the person.

11-5. Will the change in speed of a sound wave in air caused by a change of temperature have any direct influence on the frequency of vibration of sounds that originate in a room? As the temperature of the room changes performers must make adjustments in their instruments. Are these adjustments made because of the change in speed of sound in the room or something about the instrument?

11-6. Use the fact that the reverberation time is very nearly ten (9.97) times the halving time to obtain T_{rev} for the transient response of a room as it responded to sinusoidal excitation as shown below.

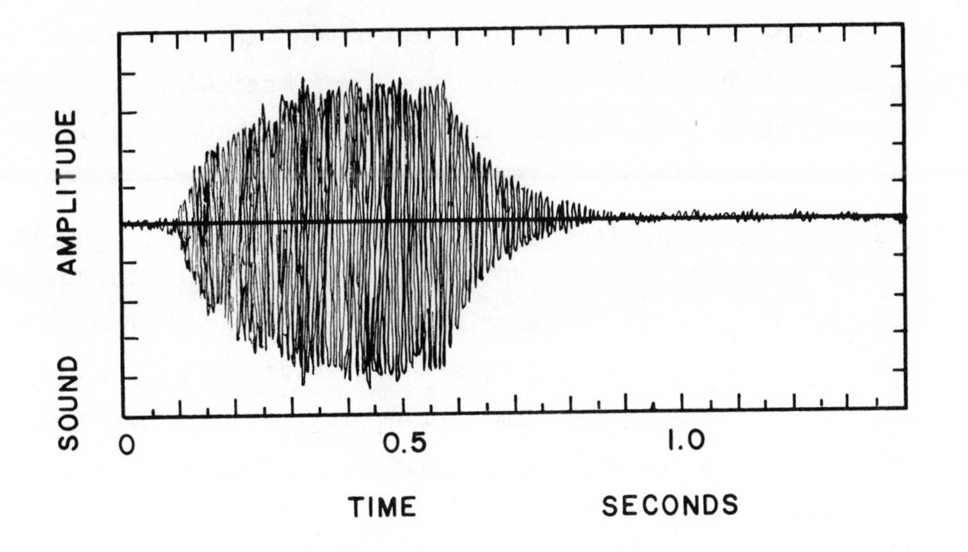

Use the relationship between the bandwidth $W_{1/2}$ and reverberation time T_{rev}

$$W_{1/2} = (3.8/T_{rev})\ \text{Hz}$$

to find the bandwidth over which the mode is strongly excited.

11-7. The relationship between the reverberation time T_{rev} and the bandwidth $W_{1/2}$ over which a mode would be strongly excited is given by the formula

$$W_{1/2} = (3.8/T_{rev})\ \text{Hz} \quad .$$

Suggestions for a proper reverberation time are often made for certain applications. Calculate the bandwidth over which a room mode would be strongly excited for each of the reverberation times T_{rev} equal to 0.040 sec, 0.1 sec, 0.5 sec, 1.0 sec, and 2.5 sec. Note that this is a very limited part of the considerations concerning room acoustics. It is useful to see the range of $W_{1/2}$ for these reverberation times.

11-8. A repetitive force can often excite a system with a steady-state response. At some particular frequency the amplitude of response can be large and have a bandwidth that depends on the damping. Such systems may act as filters for vibrational energy. In the acoustical range of vibrations, experiments have indicated that the pitch discrimination of the ear can be described in terms of the fractional changes of pitch that are detected. Assume for this problem only that the pitch discrimination data can be used as the specifications for fractional changes in frequency of an electronic filter. What sort of half width at half height does this fractional change in frequency imply for an electronic filter system that would work in the audio range and appear to pass a single frequency as far as a listener was concerned? (Note: This problem does not imply anything about how the ear works, and is included since it is occasionally a part of discussions of audio equipment. Once you have studied the various aspects of hearing you can return to this problem and comment on the possible difficulties.)

11-9. Textbooks on physics, engineering, physics of sound, etc. often contain tables or lists of the speed of sound in various materials. Such data can be found in the Handbook of Chemistry and Physics, Chemical Rubber Publishing Company; American Institute of Physics Handbook, McGraw-Hill Book Company; and other references.

a. Obtain data for the speed of sound in various gases such as air, helium, carbon dioxide; liquids such as water; and solids such as bronze, steel, iron, lead, and wood. Compare the speed in various substances relative to each other. Although one needs detailed data on coefficients of elastic behavior and density for precise comparison, your consideration of these data should reveal an interesting ranking of materials according to the speed of sound in them. Solids are not just materials that are used to form vibrating systems but are a medium that transmits sound.

b. The wave impedance can be calculated for a material if one knows the density and elasticity. The relationship is

$$z_{wave} = \sqrt{\rho \epsilon}$$

where ρ is the density and ϵ the elasticity. This relationship can be used to obtain another where the density and speed of sound are needed for the calculation, since

$$z_{wave} = \sqrt{\rho \epsilon} = \rho \sqrt{\epsilon / \rho} = \rho v$$

where v is the speed of sound. First find the wave impedance for the materials in part a. Discuss the changes in coupling between wood and steel (or iron), air and wood, and air and various metals. These comparisons are useful for a discussion latter of various musical instruments. Find data on the density and speed of sound for the fleshy portion of the ear. You may need to consult the library for books on hearing science, as for example Experiments in Hearing, Georg Von Bekesy, McGraw-Hill (1960). Discuss the coupling between the air and the ear as well as water and the ear.

12

Room Acoustics II: The Listener and the Room

12-1. Find the "half-wavelengths" in air for sounds of the following frequencies: 220 Hz (A_3), 262 Hz (C_3), 440 Hz (A_4), 587 Hz (D_5), 880 Hz (A_5), and 4186 Hz (C_8). Assume the speed of sound is 345 meters/sec.

12-2. A room measures 2.5 × 3.2 × 3.9 meters. Assume that these dimensions correspond to the half-wavelength of a low-frequency mode of the room and find the frequency of each.

Assume that the halving time for the fifth harmonic of the frequencies is 1/20 sec, and that the volume of the room is very nearly 1140 ft^3. Use the graph and information on page 178 (Fig. 11.2) of Fundamentals of Musical Acoustics by A. H. Benade on its use to estimate the number of modes excited by these frequencies. Please note that this problem is to be used for experience in finding half-wavelengths and estimating room modes. A real room of this size would normally have objects in it to absorb, reflect, and scatter sound. Reconsider the related ideas in problem 5 on page 196 of Fundamentals of Musical Acoustics.

12-3. Certain household bathrooms show some of the interesting properties of flutter echo and the acoustical problems of small rooms with hard irregular surfaces. Contrast the nature of the sound of your voice, a finger snap, or handclap when sound absorbers such as towels, bath mats, curtains, and pillows are removed and are in place. Why would these comparisons be difficult if the floor had thick wall-to-wall carpeting?

12-4. Experiment with cupping the hands at the ears in order to change hearing, and report the nature of the sounds when distinct improvement was or was not made. Compare the results of your experiment when the hands were cupped behind the ears to "gather" the sound and "focus" it into the ears with a situation where the hands cover the ears and create partially enclosed spaces. In order to produce the partially enclosed spaces each hand should be in the proper shape to grasp a large-diameter ball. Next touch your head as if to cover your ears. Sound should still be able to enter your ears by passing through the open spaces between your fingers and into the space between the hand and the ear.

a. Comment on the nature of the change in the sounds as the ears are "covered." Often a change can be noticed when you are listening in a noisy place. If another person is speaking in a noisy environment, does this procedure help in understanding his speech? Should it?

b. The results are quite different indoors than outdoors. Compare the results of both situations. Is there a difference when you use both ears as suggested or stop one ear so as to make the observations with just one ear?

12-5. It is possible to find special situations that show the excitation of room modes. Discuss the situation involving two people conversing in the hallway of a large building. The hall is eight feet wide. The walls are concrete block, and a portion of the wall adjacent to the people is a fairly large glass window. The ceiling is highly absorbent acoustical tile. The speaker notices a large increase in loudness for some words but not others. The listener, standing a short distance away, reports hearing the same effect. The speaker is a male and the effect is most evident for certain vowels.

12-6. In news articles and in course problems on the physics of sound one finds formulas for calculating the reverberation time along with discussions of possible influences of this time on acoustical uses of the room.

a. Discuss the extent to which one can use the calculation or measurement of reverberation time in evaluating acoustical uses of rooms. Limit your discussion to general classes of use of reverberation or halving time.

b. In your criticism of the use of only the reverberation time include a brief discussion of additional aspects of interior acoustics that should be included in evaluating a room for acoustical uses. In your brief development of a discussion avoid details or special situations but include clear statements of additional aspects of room acoustics.

12-7. A shell or reflecting surface is often placed behind a group of performers or a speaker. This can be done when the event is staged in the open and in an auditorium. Consider the possible acoustical effects that can arise in both situations.

a. What is the most noticeable effect when the shell is used in an outdoor situation?

b. What additional effects are possible in an indoor situation?

12-8. Consider the way in which sounds from two sources combine at a single point of measurement or observation. First consider the case where the sources are outdoors in an open space with no reflections.

a. Imagine that two speakers are placed 4 ft apart and connected to the same signal generator such that they move in phase. At a single point that is 3 ft in front of one speaker and 5 ft from the other, do the sound waves from the two speakers tend to reinforce or cancel one another?

Take the frequency to be 550/sec (Hz) and the speed of sound to be 1100 ft/sec. (Hint: First compute the wavelength; then fit in half wavelengths to see if waves are in phase or out of phase at the point of the detector.)

b. Why does a calculation of this sort have almost no meaning if applied to sounds in a room?

12-9. Some listeners may report that a particular location in a room, lecture hall, or auditorium may be a "dead spot." This is a very complicated situation where the usual simple explanation on the basis of an interference null is inadequate. Discuss reasons why the term "dead spot" may have almost nothing to do with the sound pressure level at the location in question.

12-10. Offer an explanation for the common observation that one's singing sounds different in a small enclosed space such as a bathroom or shower?

12-11. An experiment was conducted outdoors to measure the time between a pulse of sound near a microphone and a clear echo. The echo was followed by other return sounds where many reflected sounds arrived nearly together. On that day the speed of sound was 345 meters/sec. How far away was the surface that returned the clear echo?

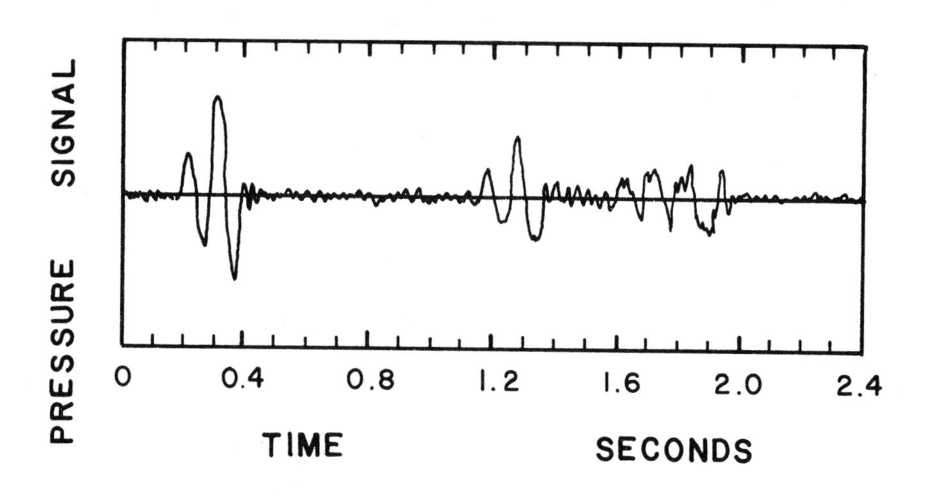

12-12. Many small "bookshelf" loudspeakers have an angular distribution of their radiation that is roughly similar to that of woodwinds and brasses. Expensive multi-element loudspeakers tend to have all-round distributions at all frequencies when the average behavior is considered. However, when the individual components of the sound are considered a single high frequency can have extremely individual and strongly directed radiation patterns. Violins and similar string instruments have a similar behavior, which is good. Why do you suppose expensive speakers could tend to distract musical listeners more than simpler ones?

12-13. In problem 5-5 the addition of two sets of cosine-type sinusoids were added to give two types of sound. One of them gave an approximation of "tick" in the form of a pulse. In order to give a single pulse one must add together a large number of sinusoids of the cosine type, all with the same initial phase. The other gave an approximation to "ssh." In order for this to work one must add together a large number of sinusoids all with random initial phases.

a. Consider how such signals would sound in a normal room. How might the ear deduce differences between the two signals, despite the fact that in a room a multiple reflected and scattered version of the sound is incident on the ear?

b. For those who enjoy mathematical computations with a hand calculator you can make a graph of "tick" with the equation

$$y(t) = \frac{1}{N} \sum_{n=1}^{N} \cos\,(1 + An)t$$

where $N = 10$ and $A = 0.2$. For the random phase version to approximate "ssh" the equation is

$$y(t) = \frac{1}{N} \sum_{n=1}^{N} \cos\,[(1 + An)t + Bn]$$

where N = 10, A = 0.2, and B = 20. Use a hand calculator and make a graph of these two relationships for the values given for the parameters A, B, and N.

13

The Loudness of Single and Combined Sounds

13-1. Doubling the amplitude of a single frequency source has the effect of adding 6 dB to the sound pressure level, whereas doubling the number of identical sources only adds 3 dB to the sound pressure level. Verify this statement and comment on the loudness for both cases when the starting point is an 80 dB source at 250 Hz. What is the loudness of this sound?

a. Find the loudness if the amplitude of the 80 dB source is doubled.

b. What is the loudness if two identical sources at 250 Hz and 80 dB each are sounded together?

13-2. Doubling the amplitude of a single frequency source has the effect of adding 6 dB to the sound pressure level, whereas doubling the number of identical sources, say in a room, only adds 3 dB to the sound pressure level. Verify this statement and comment on the loudness for both cases when the starting point is an 80 dB source at 250 Hz. What is the loudness of this sound?

13-3. Speculate on the possible meaning of the term "penetrating tone" as used by musicians in describing certain types of musical sounds. Develop a brief description of the term based on the point-of-view of acoustical perception and the physical nature of the sound spectrum. Assume that the tone is a characteristically musical one consisting of a group of sinusoidal harmonic components and estimate the relative sound pressure levels for the harmonics.

13-4. a. Compare the perceived loudness of a pair of sinusoids whose frequencies are 60 Hz and 500 Hz, assuming they have the same sound pressure level (SPL) of 72 dB.

b. Consider the situation where each of these frequencies are now the fundamental component of a set of harmonics. In each case the first three harmonics have equal amplitudes: $a_1 = a_2 = a_3$. The higher harmonic components have amplitudes $a_4 = a_1/2$, $a_5 = a_1/4$, and $a_6 = a_1/8$, so that the SPL's are 66, 60, and 54 dB, respectively. There are no components with higher frequencies. In order to make this comparison the SPL's of the components need to be evaluated. Suppose that the fundamental frequency for both sounds (60 and 500 Hz) is present with an SPL of 72 dB. Adapt the data in Figure 13.2 of _Fundamentals of Musical Acoustics_ by A. H. Benade so as to confirm the stated SPL's of all the higher harmonic components of the two sounds. (If you decide to use a hand calculator for this task give one worked example of the equations used.) Find the loudness of these two tones.

c. Compare the loudness found for parts a and b. Give the incremental increase in loudness when the harmonics are supplied.

13-5. Consider an electronic source of "white noise" (the rushing sound of the kind one gets between channels of a TV or FM set) combined with a 200 Hz sinusoid. When one listens to the combination with headphones it is possible for the noise to mask the sinusoid when it is set at a sound pressure level of 70 dB. However, when signals that produce the same sound pressure levels for the noise and the sinusoid are fed to a single loudspeaker in a room the sinusoid is audible. Discuss the reasons for this observation about the two listening conditions.

13-6. For a sound made up of sinusoids whose frequencies are whole-number multiples of one another (forming a harmonic series) it is quite safe to add loudnesses for the different components of sound when the masking effects are drastically reduced as one listens to them in a room. In wind instruments the typical behavior is for the sound produced at the pianissimo (pp) level to be very nearly a single sinusoidal at the playing frequency. As one plays louder, new harmonically-related components are generated. These come in one-by-one, and grow in a well-defined way. The loudness implications of this typical behavior can be introduced very conveniently via the following simplified numerical example. Suppose that the instrument when played pp generates a sinusoid having a sound pressure level (SPL) of 50 dB at 250 Hz (at the listener's ear). Assume that when the player is sounding his instrument at the fortissimo (ff) dynamic level he is generating the 250 Hz component at 80 dB (1000 fold increase in actual sound intensity). He is also producing at the same time a 500 Hz component (second harmonic) at 77 dB (one-half the sound intensity at 250 Hz) and a 750 Hz component (third harmonic) at 74 dB (one-fourth the sound intensity at 250 Hz). We will pretend that he is generating only negligible amounts of sound intensity at any other frequency.

a. Compute the loudness of the sound produced by our musician when he is playing pp, and when he is playing ff.

b. Compute the SPL that would be required for a 250 Hz simple sinusoid if it were equal in loudness to the musical instrument played ff. (Hint: See page 252, example 5, of _Fundamentals of Musical Acoustics_ by A. H. Benade.)

13-7. By what factor will the loudness increase if the sound pressure level is increased from 70 dB to 73 dB (doubling the intensity) at 200 Hz?

13-8. A B_b clarinet is played in such a manner as to produce a tone whose fundamental component is close to 250 Hz. What note is being played? A listener in the room hears the tone, with the sound pressure level (SPL) distributed among the harmonics as follows:

1st Harmonic	70 dB
2nd Harmonic	40 dB
3rd Harmonic	60 dB
4th Harmonic	40 dB

where the SPL's are measured relative to the zero dB reference level at threshold.

a. Compute the loudness of this combined sound in a room where the effect of masking is obscured so that the loudness is the combination of the loudness of the individual components.

b. Imagine another acoustical environment where the loudness of a musical tone can be fairly well estimated by adding the loudness of the fundamental component to the resultant combined loudnesses of all the other harmonics. What is the loudness in this situation?

c. For this clarinet tone the loudness will not be badly estimated from the sum of the loudnesses of the first and third harmonics. Explain this assertion by comparing the loudness computed by the methods in a and b with a calculation for the combination of only the first and third harmonics.

d. Explain a further assertion that a woodwind adjusted so that the first prominent harmonic above the fundamental is strong compared to the fundamental will generally be considered a loud instrument, even though a laboratory measurement may not give a particularly strong value of the sound power.

13-9. Consider six separate sinusoids which are presented one-by-one to a person with normal hearing. The frequencies are 50, 100, 200, 500, 2000, 10,000 Hz.

a. Rank these sounds in order of loudness for the case where each is presented with a sound pressure level of 70 dB.

b. Re-rank these same sinusoids for the case where their sound pressure levels are 100 dB.

13-10. Consider a situation where you are supplied with four sinusoidal components of a sound whose amplitudes and frequencies can be freely selected by any criteria you wish. (Note: The frequencies need not be harmonically related.) Given that one of these components has an amplitude of 60 dB sound pressure level, select the amplitudes and frequencies to give the maximum net loudness for the composite sound. Explain your line of reasoning. Note that there are several correct solutions to this problem but you should only work out the amplitudes and frequencies for one. Why should the components of the sound not be too closely spaced?

13-11. One can conduct some pencil and paper experiments on the loudness of various sounds. Imagine that one is listening to a tone with a set of harmonically related partials in a room with ordinary acoustical properties. The tone has the harmonically related partials 880, 1320, 1760, 2200, 2640, and 3080 Hz. For the purpose of experimenting with the procedures for combining sounds to obtain loudness, the listener is assumed to have a threshold that indicates a loss in the high frequency portion of the hearing range.

80 Each of these partials has a pressure amplitude corresponding to a sound pressure level of 20 dB.

a. Use the figure below to decide on the audibility of each of the partials considered by itself.

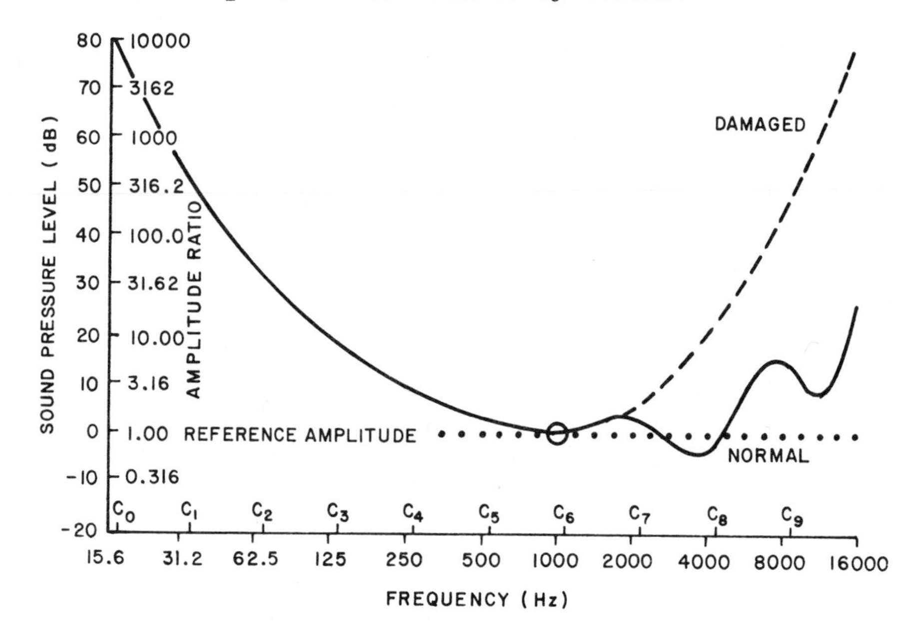

Figure 13.1. A fictitious hearing loss threshold curve. (Adapted from Fundamentals of Musical Acoustics, Arthur H. Benade, Oxford University Press, 1976 with permission.)

b. By means of rules developed for assigning pitch to a collection of harmonically related partials (Sections 5.6 and 5.7 of Fundamentals of Musical Acoustics by A. H. Benade) and the relationship between named musical notes and frequency (Figure 2.1 of Fundamentals of Musical Acoustics) decide the note name one would give to such a very soft sound. Does this particular threshold curve present any problems in pitch assignment compared to a normal curve?

13-12. Consider the six separate sinusoids discussed in problem 13-9 for a person with normal hearing in the situation where they now are presented one-by-one to a person with the hearing loss shown in Figure 13.1. The frequencies are 50, 100, 200, 500, 2000, and 10,000 Hz with a 70 dB sound pressure level. Rank these sounds in order of loudness and compare this ranking with those found in part a of problem 13-9.

13-13. Consider the following situation: You are sitting alone in a deserted music hall when a lone violinist comes in and plays loudly enough that at your ear the sound pressure level is about 50 dB above threshold. In the next few minutes another player appears, followed by two more, and then by four more, etc., doubling the number each time until a total of 32 violinists are on stage. Assume that they are all playing the same part, and are all playing so as to make equal contributions to the sound in the room. That is, each one playing by himself will produce a 50 dB level at your ear.

a. Calculate the perceived loudness of the sounds produced by the players as they drift in and begin to play.

b. See if the total range in loudness (difference for one player) is the same if all players radiated sound at a level of 60 dB instead of 50 dB. Comment on the practical implication of this for the conductor and composer.

13-14. I wish to point out at the start of this problem that the same ideas are found in Fundamentals of Musical Acoustics by A. H. Benade, Section 13.5 starting on page 234. You could in fact use the curves found there to do this example. I believe it will be instructive to complete some graphs for yourself to illustrate the way sounds combine.

The loudness of sounds can be classified in an approximate way into the groups soft, ordinary loudness, and great loudness. For sounds that are louder than the soft range, graphs can be made of the relationship between the combined loudness of the sounds and the number of identical sounds.

82 The first graph is to contain the often-quoted observation that ten violins all playing mezzo-forte sound twice as loud as one violin playing at this loudness. This statement gives two data points on the "log-log" graph: the loudness of one violin and the loudness of ten violins. Complete the graph of sound power in a room as it depends on the number of players by drawing a straight line through these two points. This is a non-trivial approximation that you can verify by consulting Figure 13.5 of Fundamentals of Musical Acoustics by A. H. Benade. The graph you have prepared can be used to find the approximate loudness of any number of violins.

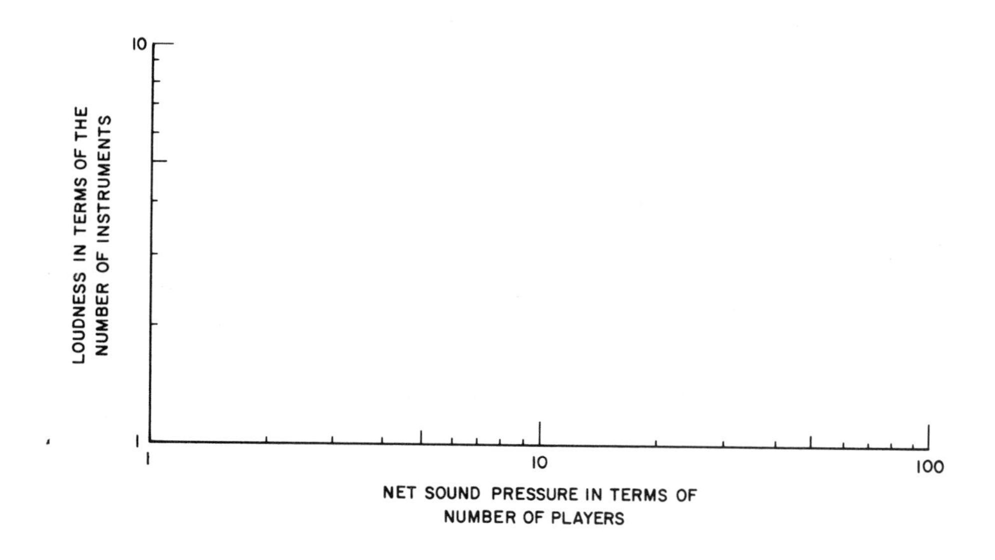

In a second graph you are to make a plot with linear coordinates of these original two points and some additional ones by transferring values from the "log-log" to the linear graph. From a "log-log" graph you cannot deduce the obvious fact that zero violins produce zero sound but you can indicate this observation on the linear graph.

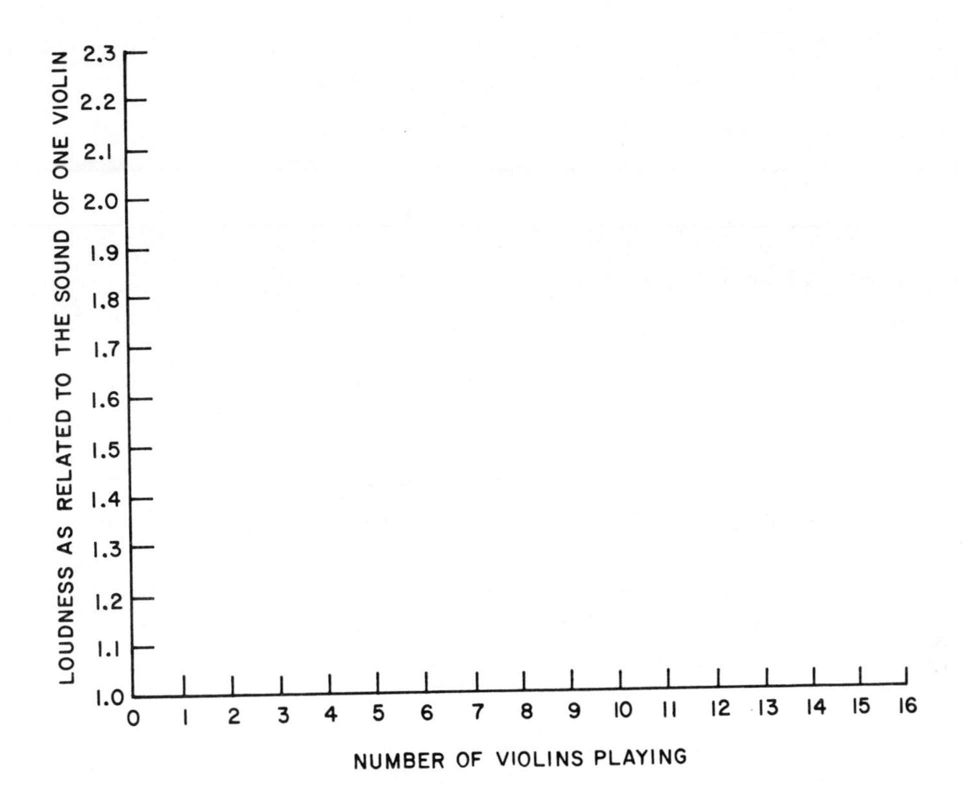

a. Transfer the data points that represent the loudness of 2, 4, 3, 14, 16, and 20 violins in terms of one violin. As an intermediate step make a list of these data points. Connect these points and add any others you need to sketch a smooth curve that passes through all the points.

b. Comment on the relative change in loudness as additional performers are added to the group when it is limited to two or three performers compared to adding one player to a group of ten.

c. Use the graph to discover approximately how many violins would be required to obtain three times the loudness of one violin.

d. Comment on the combined loudness of a group of 76 trombones compared to a single trombone when it is assumed that the loudness of combined trombones obeys the rules of identical sources as exemplified by violins.

14

The Acoustical Phenomena Governing the Musical Relationships of Pitch

14-1. It is possible to illustrate the effect of a linear transfer function graphically. The diagram below shows how several points of the input curve are transferred by projection onto the transfer function and then to the output curve. Complete the graph of the output and compare the shape of the input and output curves. The input is a a sinusoid. Divide the input and output axes into finer

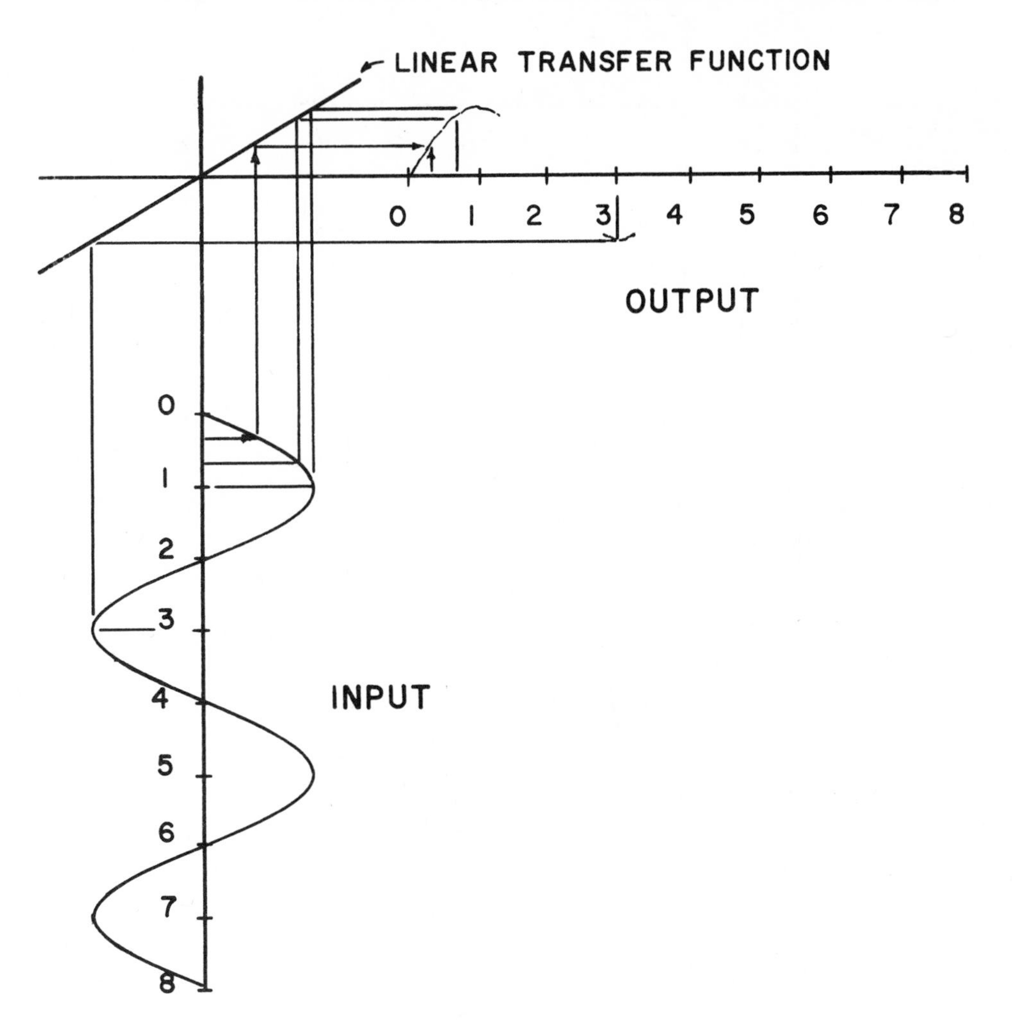

 divisions than the eight that are shown. Individually project each corresponding amplitude upward to the transfer function, and then horizontally to the corresponding position along the output axes where it forms the amplitude of the output curve. The output curve would represent the response to the input curve as the excitation or driver. In a linear system equal increments of input (stimulus) result in equal increments of output (response). Confirm that the linear transfer function in this example has such a property.

Many systems do not have linear transfer functions. As a mathematical example of a nonlinear transfer response

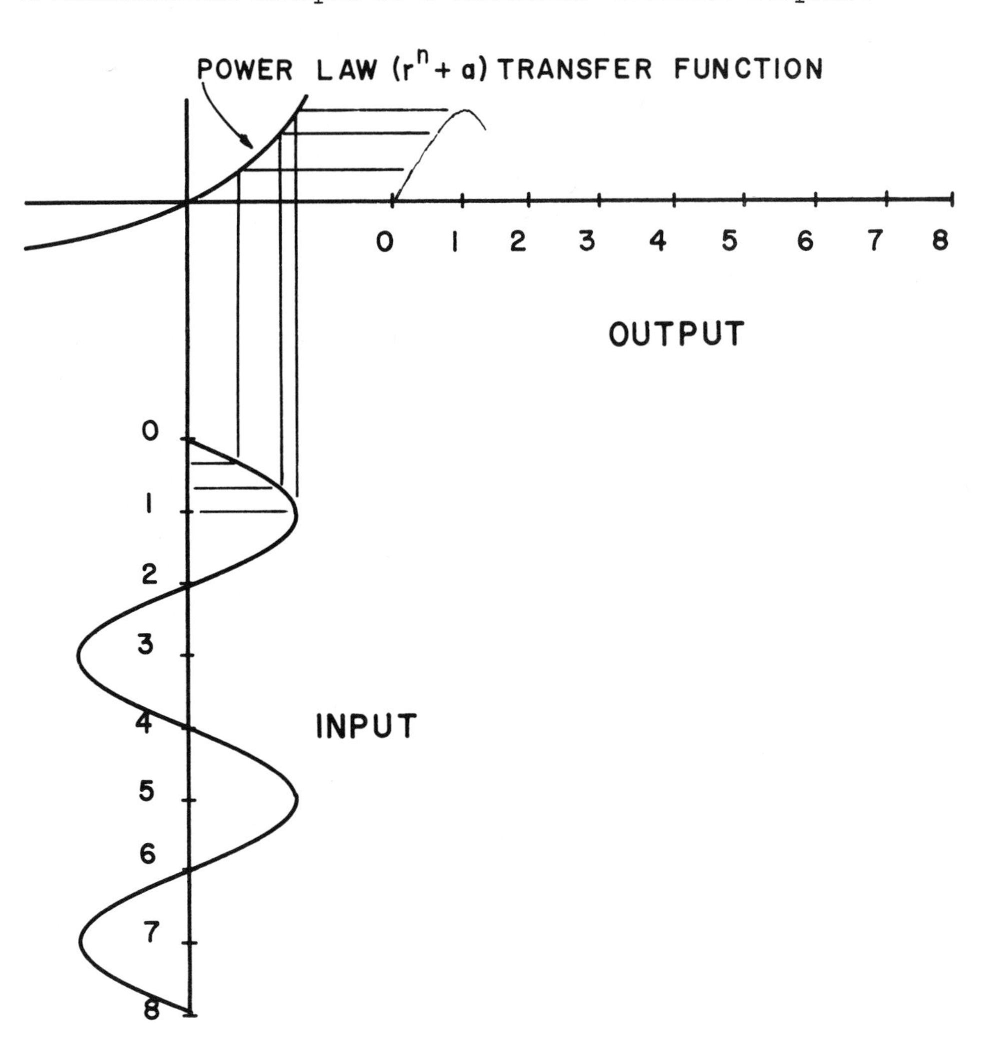

function consider a "power-law," r^n function. Complete the graph of the output and comment on the shape of the input and output curves. The input is a sinusoid. In a nonlinear system equal increments of input (stimulus) result in unequal increments of output (response). Confirm that the power law transfer function has such a property.

Transfer functions such as $y = x^2$, $y = x^3$, and $y = e^x - 1$ could be used to describe other nonlinear responses to sinusoidal drivers. Some response functions are not expressible as mathematical functions. Three examples are graphed below. Discuss the response of these functions (by inspection or with a simple sketch) for small amplitude and large amplitude sinusoids. (For these diagrams an amplitude of 1/2 cm or 1/4 inch would be small. Amplitudes greater than 1 cm or 1/2 inch would be large.) Comment on the general characteristics of the responses for the two conditions for the input signal.

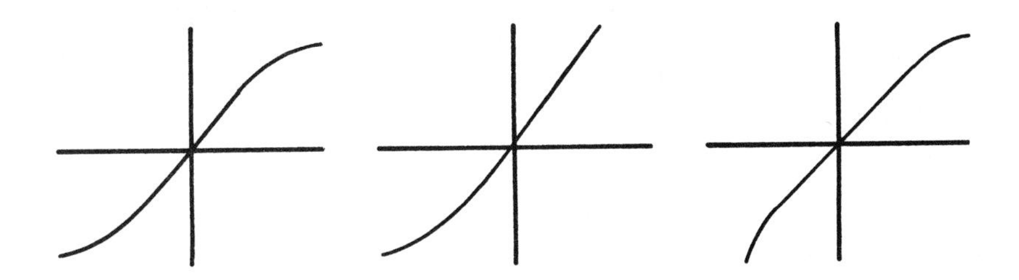

14-2. When a sound consists of a set of harmonic partials a strong sense of pitch results even though several of the components are missing. The same nonlinear production of heterodyne components can be used to describe how a collection of almost-harmonic components can give a strong impression of pitch while sounding somewhat rough. Carry out the calculation of heterodyne components for the three almost-harmonic components.

200, 306, 497 Hz

Arrange your data so the heterodyne components group into "clumps" of nearly the same value. Since the pitch of a beating pair of sinusoids often lies about midway between the pitches of the two separated sinusoids, you might replace each "clump" by its own average frequency. Notice how well marked the pattern has become and how it is a quasi-harmonic set with a strong resemblance to a harmonic set. The beats within each clump contribute to the roughness of the sound.

14-3. The procedure for using heterodyne components to find the pitch of almost-harmonic components can be applied to a set of almost-harmonics consisting of more than a couple of members. Carry out the calculation of heterodyne components for the four almost-harmonic components

100, 202, 297, 410 Hz

Arrange your data into clumps where you have indicated the quasi-harmonic nature of the set. Find the average of each clump and note the beat rates. Include several examples of heterodyne components that are grouped with the 410 Hz component. This component was selected to show that a large departure from harmonicity in the original set will give heterodyne components that are not in well-defined clumps. Verify that this component contributes to a larger beat rate than the other members of the set. If there is too much inharmonicity, the heterodyne components splatter all over the place and leave no impression of orderliness.

14-4. Consider what happens if two instruments are set to play together; one has a recipe with components oscillating at the rate of 400, 800, 1200, 1600, and 2000 Hz while the other one has a recipe with components at 301, 602, 903, 1204, and 1505 Hz.

a. Work out the heterodyne components belonging to the combined sound for frequencies up to 2000 Hz. You will find that these components group themselves into small clumps.

b. Make a list of these heterodyne components along with their parents in tabular form that will clearly indicate where we can expect to hear beats between closely-spaced components. In other words, gather everything into small clumps, each of which contain similar frequencies.

c. Compare the number of clumps you found and the number in each clump with those that would occur for a similar calculation involving the intervals of an octave or a musical fifth.

14-5. Consider what happens when you search for clumps of heterodyne components from the three components

200, 327, 382 Hz

Can you find any kind of arrangement that forms clumps so that a single pitch might arise from the components? (The situation is clear for components with frequencies less than 1000 Hz.) Speculate on what one would hear for these components.

14-6. Suppose that a woodwind player is trying to tune to a standard A-440 tuning fork. At some instant in the beginning of his blowing, his instrument is generating a tone consisting of the following partials:

441, 882, 1323, 1764 Hz

Assume that the tuning fork is producing its tuning partial. What is the heterodyne tone produced by the ear? Consider only the simplest components and list the beats that form the basis for a judgment of the tuning error.

14-7. Consider two tones P that consists of 400, 800, and 1200 Hz and Q that consists of 501, 1002, and 1503 Hz in a situation where they are sounded together.

a. Find the clumps of heterodyne components for the collection of original components and the simplest heterodynes. Do not extend the calculation above 1503 Hz.

b. Find the additional components produced if tone P includes the fourth harmonic 1600 Hz and tone Q includes the fourth component 2004 Hz. Also, make a tabulation of the combined results of parts a and b. Comment on the degree that the musical interval between P and Q is defined in the two cases.

c. What single pitch is defined by the components of tone P alone; by the components of tone Q alone; by the combined components of tones P and Q? Speculate on possible reasons for not always hearing the latter case.

14-8. Given a sound made up of four unrelated components, choose a set of frequencies which would combine to produce the most unmusical, dissonant, rough sound that it is possible for you to devise. Explain the reasons for your choice and list the "rules" you used to create this "beautiful noise." (Note: You do not have to stick to any of the ordinary tunings for these tones; they can have any frequencies you choose for the components.)

14-9. Here is a different version of problem 14-8. Given a tone made up of four harmonically-related partials and having a fundamental frequency of 400 Hz, choose the fundamental frequencies of two similar tones which would go along with the first to produce the most unmusical, dissonant, rough sound that it is possible for you to devise. Explain the reasons for your choice and list the "rules" you used to create this "beautiful noise." (Note: You are free to choose any frequencies you wish for the fundamentals.) Suppose that you are now given a total of four tones to combine with the same specifications. Would you simply choose a fourth to go along with the original three, or would you come up with a newly-adjusted set of sounds?

14-10. Although this problem appears to be separable into two questions, the contrast between the conclusions drawn in each part is an important part of the task. Read through the entire problem and identify the similarities and differences before the details of each part are considered.

Consider two trains of pulses where one has a slightly greater repetition rate than the other. These are both fed into the same loudspeaker so that in the course of time the output signal has the appearance shown below when viewed on an oscilloscope.

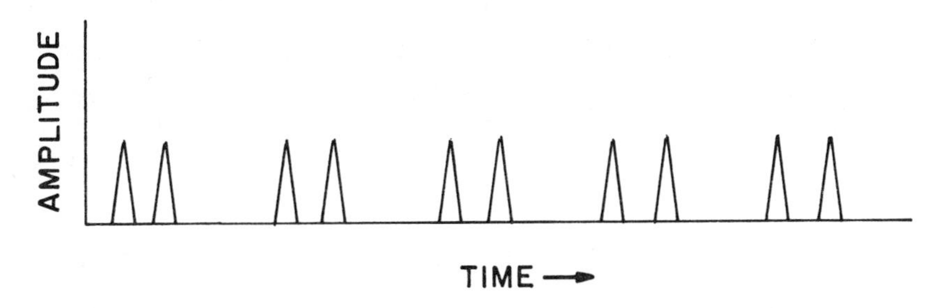

Since the two frequencies are slightly different for the sets of pulses, one will slowly slide over relative to the other. The appearance of the pulses later will then be as shown below.

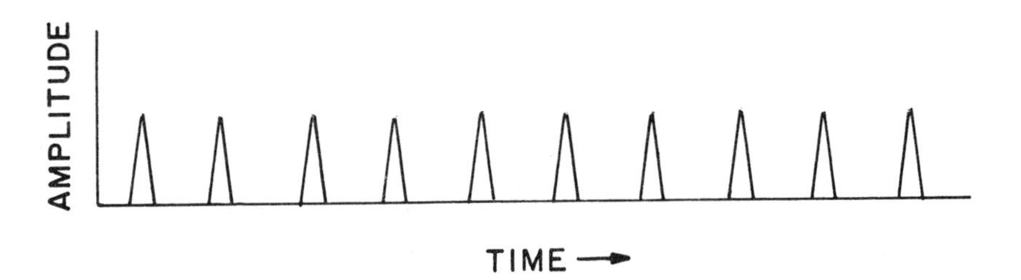

 The frequencies are very nearly the same so that the peaks are almost centered for a short time before the pattern returns to that shown first.

a. Suppose the repetition rate of one train of pulses is 250/sec and the other is 250.5/sec. Describe the perceived pitch of the sound as time goes on. Also, describe the loudness of the sound as time goes on.

b. Suppose that instead of the pulses in part a the loudspeaker is now supplied with a pair of sinusoids with repetition rates of 250/sec (Hz) and 250.5/sec (Hz). What is the pitch of the sound? Describe the variation of loudness of the sound as time goes on.

c. Although the two situations seem similar, the conclusions reached in the two cases above are drastically different. A pulse train consists of a set of harmonic components. Use this property of a train of pulses to explain the different conclusions in the two situations.

14-11. Consider two repeating signals where one has a slightly greater repetition rate than the other. The wave form of each is a triangular wave similar to the one shown here.

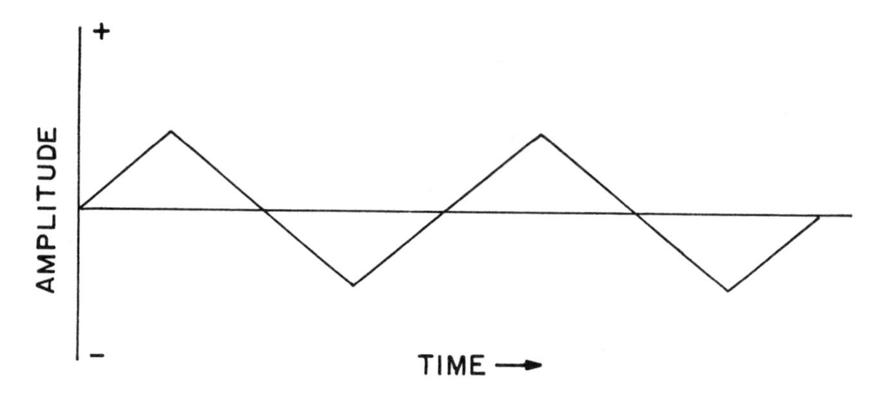

These are both fed to the same loudspeaker in a situation like that of problem 14-10 above. Suppose the repetition rate of one triangular wave is 250/sec (Hz) and the other is

250.5/sec (Hz). Describe the perceived pitch of the sound as time goes on. Describe the loudness of the sound as time goes on. While a pulse train is made up of a set of all the harmonic components, a triangular wave consists of a set of the odd frequency harmonic components. Use this property of the wave shape to explain the conclusions of this situation as compared to problem 14-10.

15

Successive Tones: Reverberations, Melodic Relationships, and Musical Scales

15-1. Consider the beats produced by a slightly mistuned or "almost" fifth. One sound P has the components 200 and 400 Hz and the other sound Q has the components 301 and 602 Hz. Find the heterodyne components lying below 1002 Hz of the two sources sounding together for values less than 1002 Hz. Group them into clusters ordered according to increasing frequency values. A tabular form with cluster values in columns is a convenient format. Next add a 600 Hz component to sound P and find the additional heterodyne components. Enter them in the list of the appropriate cluster. Comment on the increase of components in each cluster and determine the beat rate. Next add a third component of 903 Hz to sound Q and find the additional heterodyne components. Verify some properties of this slightly mistuned fifth. Both sounds are made up of harmonic partials. The beat rate for each cluster of heterodyne components is the same. The most frequent beat rate is 2 Hz. What other beat rates are possible, and what relationship do they have to each other?

15-2. Consider what happens if two instruments are played together, one of which has a musical tone made up of components oscillating at the rate of 400, 800, and 1200 Hz, while the other one produces components at 501, 1002, and 1503 Hz.

a. List these components in order of increasing frequency and use the rules for pitch determination to assign a pitch to this composite tone.

b. Find the heterodyne components belonging to the combined sound. Limit your list to frequencies of 1500 Hz or less. Your list should show that these components are distributed in small clumps or groups with nearly the same frequency.

c. Make a list of these heterodyne components along with their parents in some sort of tabular form that will clearly indicate where we can expect to hear beats between closely-spaced components.

15-3. One form of imperfection that is particularly troublesome in some Hi-Fi sets occurs when a sinusoidal input from the record is not radiated as a sinusoid by the loudspeaker. One finds that slightly mistuned musical intervals sound immensely rougher when played through such an imperfect outfit than when they are listened to directly from the players or via a perfect playback outfit. Account for this effect in terms of the properties of the ear and the analogous distortions of the loudspeaker. You could consider the tones

P: 200, 400, 600 Hz
Q: 302, 604, 906 Hz

for calculating a sufficient number of heterodyne components to demonstrate the effect. Verify that the increased roughness results from heterodyning for both the loudspeaker and the ear.

15-4. Consider yourself to be part of a superb woodwind trio, given the following opening notes to a composition:

You can recognize this as a sequence of two identical minor sixths and two major thirds. Recall that the 5/4 frequency ratio gives a well-marked beatless major third, and that an 8/5 frequency ratio gives a recognizable message of perfect tuning as a minor sixth. In order to simplify calculations assume the lowest note in our composition to have a fundamental frequency of 400/sec (Hz). Let it have three harmonic partials that are of sufficient strength to matter. Let the designation of this note be P just to prevent any prior ideas of musical tones from leading to confusion. Compute the frequency for the second-from-the-lowest note if it is a perfect third above P and call it Q. Compute the frequency of the top note as if it were a perfect third above Q and call it R. Next compute R as if it were a perfect minor sixth above P. Now describe the results when the trio performs the composition. What would be a reasonable course to follow if this were actually performed?

15-5. Now look into the properties of a beatless major triad, which can be viewed as the combination of a beatless fifth (frequency ratio 3/2) and a beatless major third (frequency ratio 5/4). Then the three tones are

P: 400, 800, 1200 Hz
Q: 500, 1000, 1500 Hz
R: 600, 1200, 1800 Hz

a. Find the heterodyne components for pairs of these tones, and find the number of components for the frequencies 100, 200, 300, ... 1800 Hz.

b. Repeat the problem when the tone Q is based on 480 Hz, which converts the sound to a minor triad. Comment on any musical implications that may be noticed.

15-6. It is a part of any discussion of the action of the ear in producing heterodyne tones to include the type known as combination tones. The type to be considered are those combination tones corresponding to the differences as they result in a new harmonic series. This series can have a pitch that is often part of our musical awareness. This situation often appears in music. Search through some written music or listen to records to find your own examples. Some starter examples are the English horn and clarinet in opening bars of Largo from Dvorak's New World Symphony; recapitulazione della prima parte, 3rd movement Bartok Quartet 3, 1927 (they will be scattered); and Hindemith Mass 1963 in the Gloria. Look for things such as

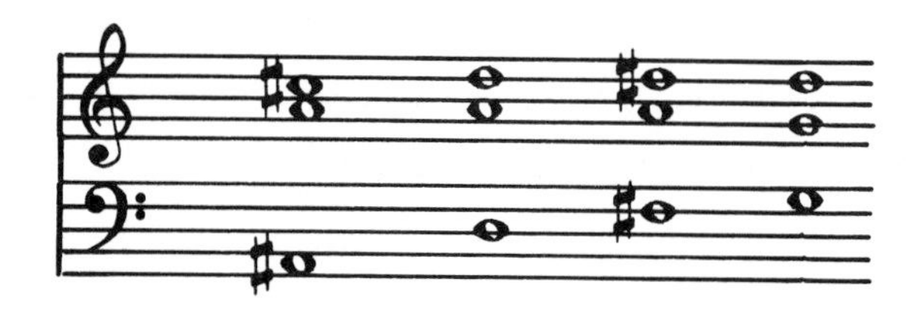

In the score one can also find that the voice or cello, for example, is also on top of this lowest tone. These are often quite hard to carry off successfully even for expert musicians. Composers such as Richard Strauss, Berlioz, and Beethoven are good sources but examples are to be found everywhere.

15-7. You are standing close to two singers who have strong, steady (vibrato-free) voices under perfect control of pitch. One of these singers produces the sustained tone D_4, while the other produces a G_4-to-G_4 trill, the lower member of the trill being given a tuning that is a beatless fourth of G (frequency ratio of 4/3) relative to the D. You will hear an additional low component that accompanies the tones from the singers. This component will also show a trill. There is something peculiar about this trill. Find out what is going on, explain it clearly from the point of view of physics and perception, and write out the whole phenomenon as clearly as possible on the musical staff.

15-8. Consider the implications of the musicians' preference for using a vibrato as it may bear on the question of identifying the tone color of a given instrumental sound (or "voice") as it is heard in a room. Be as quantitative as possible in describing the way in which vibrato can be useful to the enhancement of a desirable tone color. Include a few examples of unhelpful things that can happen to the tone color as well.

16

Keyboard Temperaments and Tuning: Organ, Harpsichord, Piano

16-1. In order to facilitate the computation of frequency ratios and the deviations of frequencies relative to equal temperament, some data in tabular form are presented in Tables 16.1, 16.2, and 16.3. Figure 16.1 shows a standard piano keyboard with the customary designation of key number. Indicated in the G and F clefs are the range of C's that encompass the useful range of musical instruments. A number of schemes for identifying the named notes have been used in the past and are indicated in the figure. A convenient scheme for use with a typewriter is to indicate the octave with a subscript and to designate all the notes in capital letters with the mid-space note as C_4 (middle C).

Table 16.1 gives the frequency ratios of the successive semitone steps in the equal temperament octave. This table can be used to determine the frequency for any named note

Table 16.1

Frequency Ratios of the Equal Temperament Octave

Cents	Ratio	Cents	Ratio
1200	2.00000000/1	600	1.41421356/1
1100	1.88774863/1	500	1.33483985/1
1000	1.78179744/1	400	1.25992105/1
900	1.68179283/1	300	1.18920711/1
800	1.58740105/1	200	1.12246205/1
700	1.49830708/1	100	1.05946309/1

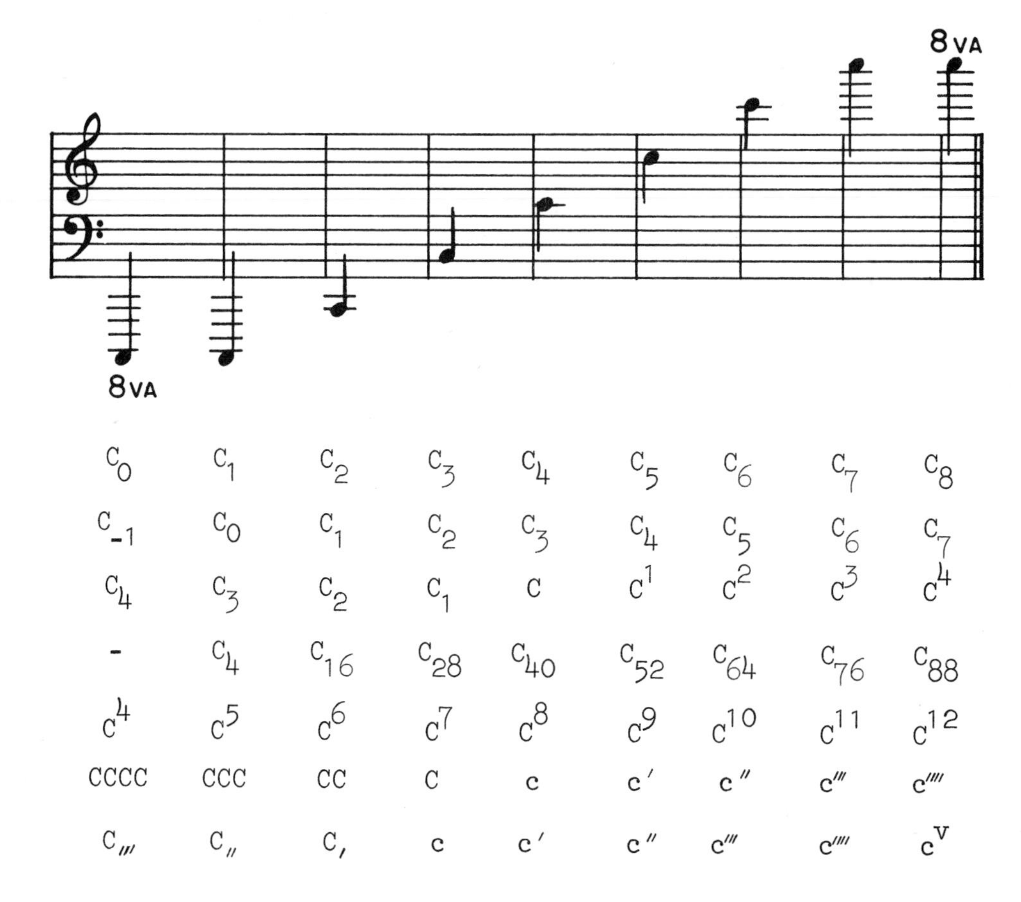

C_0	C_1	C_2	C_3	C_4	C_5	C_6	C_7	C_8
C_{-1}	C_0	C_1	C_2	C_3	C_4	C_5	C_6	C_7
C_4	C_3	C_2	C_1	C	c^1	c^2	c^3	c^4
-	C_4	C_{16}	C_{28}	C_{40}	C_{52}	C_{64}	C_{76}	C_{88}
c^4	c^5	c^6	c^7	c^8	c^9	c^{10}	c^{11}	c^{12}
CCCC	CCC	CC	C	c	c'	c''	c'''	c''''
$C_{'''}$	$C_{''}$	$C_{'}$	c	c'	c''	c'''	c''''	c^{V}

Figure 16.1. The top part of the illustration shows a drawing of a standard piano keyboard with the customary designation of key number. The middle part shows the location on the G and F clefs of the C's that encompass the useful range of named notes of musical instruments. The bottom part shows a number of methods that have been used for indicating in staveless notation the C's directly above. The first method in the list is used in this problems book.

---

in equal temperament. Recall that the sequence of note names is C, $C^{\#}$, D, $D^{\#}$, E, F, $F^{\#}$, G, $G^{\#}$, A, $A^{\#}$, B. The standard reference frequency for A_4 is 440 Hz. Each interval between the equal temperament notes is 100 cents. Those frequencies within an octave above A_4 are found by multiplying 440 Hz by the value of the ratio in Table 16.1, and for those within an octave below A_4 by dividing by the value of the ratio. To extend the frequencies beyond the octave one multiplies or divides by the proper number of factors of two. For example, to find the frequency of C_5 the computation is

$$(440 \times 1.18920711) = 523.2511284 \text{ Hz}$$

which is 523.25 Hz when rounded to two decimal places (enough accuracy for most purposes). (Note: C_5 is not a full octave above A_4.) The frequency of B_2 could be found from the following computation:

First find B_4: $440 \times 1.12246205 = 493.8833020$ Hz

Then find B_2: $493.8833020/4 = 123.4708255$ Hz

which can be rounded to 123.47 Hz. One could use other combinations of the octave and semitone interval for this computation.

a. Compute the equal temperament frequencies for these named notes: A_1, $D_2^{\#}$, C_4, G_6, $C_7^{\#}$.

b. The lowest note on a standard piano is A_0 and the highest note is C_8. What are these frequencies?

16-2. The frequency intervals in the idealized "just scales" are expressed as the ratio of whole numbers. A list of some of these just frequency relationships is presented in Table 16.2. The procedure to find the desired frequencies is to multiply or divide by the appropriate ratio. Suppose the equal temperament C_4 is selected as the reference note (the tonic) which is 261.63 Hz. The note that is a minor seventh above this has a frequency of

$$(9/5) \times (261.63) = 470.93 \text{ Hz}$$

and if one were to calculate frequency for the same "just

Table 16.2

Some Just Frequency Relationships as Ratios and Cents From a Starting Note

Interval Name	Ratio	Cents	
Octave	2/1	1200	1200.00
Major seventh	15/8	1088	1088.27
Minor seventh	9/5	1018	1017.60
Grave minor seventh	16/9	996	996.09
Harmonic minor seventh	7/4	969	996.09
Major sixth	5/3	884	884.36
Minor sixth	8/5	814	813.69
Perfect fifth	3/2	702	701.96
Diminished fifth	64/45	610	609.78
Augmented fourth	45/32	590	590.22
Perfect fourth	4/3	498	498.04
Major third	5/4	386	386.31
Minor third	6/5	316	315.64
Major second	9/8	204	203.91
Minor second	10/9	182	182.40
Semitone	16/15	112	111.73
Unison	1/1	0	0.00

note" an octave higher it would be 941.86 Hz. In order to find the frequency a major third below C_4 the calculation would be

$$(4/5) \times (261.63) = 209.30 \text{ Hz}$$

and if one were to calculate the same "just note" two octaves lower it would be 52.33 Hz.

a. Calculate the frequencies for the notes in a "just major scale" starting with the equal temperament C_4 for the intervals of a major second, major third, perfect fourth, perfect fifth, major sixth, major seventh, and the octave.

b. Calculate the frequencies for the notes in a "just minor scale" starting with the equal temperament C_4 for the intervals of a major second, minor third, perfect fourth, perfect fifth, minor sixth, minor seventh, and the octave.

16-3. In numerical problems that involve the frequency of vibration of various physical systems and musical instruments it is not uncommon to find frequencies that are not in exact agreement with those of equal temperament. The division of the octave into 1200 cents results in a convenient way of referring any frequency to its nearest equal temperament named note. Table 16.3 presents the necessary data for the octave C_4 to C_5. In order to use this table consider first the frequency 436.1 Hz. The closest value of frequency is 436.20 Hz, and the frequency 436.1 Hz can be expressed as A_5 - 15 cents. (To use fractions of a cent is entirely too fine a distinction for this problem.) Next one might find that the frequency of vibration of the first mode of a taut string is 875 Hz. When this frequency is divided by two the result is 436.5 Hz; hence the frequency of 875 Hz can be expressed as A_5 - 10 cents. (Why is it not A_5 - 20 cents?)

a. Find the nearest named notes in equal temperament and the deviations (minus is flat and plus is sharp) corresponding to the frequencies of repetition of 60 Hz, 100 Hz, 120 Hz, 500 Hz, 550 Hz, 1000 Hz, 1568 Hz, 3520 Hz, and 7500 Hz.

b. Find the nearest equal temperament named note and the deviation for the "just scale" frequencies obtained in problem 16-2.

Table 16.3

A Listing of Frequencies by One-Cent Increments that Correspond to Musical Notes in the Octave Beginning with C_4 for Equal Temperament — The columns represent the twelve semitone intervals in the scale of equal temperament. Progression down a column represents an increase in frequency or sharpness in cents relative to the head of the column. At the foot of each column is the named note designation of the last entry. Progression up a column represents a decrease in frequency or flattening in cents relative to the named note entered at the foot of the column. A tone can be specified either as the repetition rate (frequency) in hertz or a named note with the deviation in cents (+ if sharp or - if flat). When it is appropriate a frequency can be specified as a named note and deviation in cents.

CENTS SHARP	C_4	$C^\sharp_4$	D_4	$D^\sharp_4$	E_4	F_4	$F^\sharp_4$	G_4	$G^\sharp_4$	A_4	$A^\sharp_4$	B_4	
0	261.63	277.18	293.66	311.13	329.63	349.23	369.99	392.00	415.30	440.00	466.16	493.88	-100
1	261.78	277.34	293.83	311.31	329.82	349.43	370.21	392.22	415.54	440.25	466.43	494.17	99
2	261.93	277.50	294.00	311.49	330.01	349.63	370.42	392.45	415.78	440.51	466.70	494.45	98
3	262.08	277.66	294.17	311.67	330.20	349.83	370.64	392.68	416.02	440.76	466.97	494.74	97
4	262.23	277.82	294.34	311.85	330.39	350.04	370.85	392.90	416.27	441.02	467.24	495.03	96
+5	262.38	277.98	294.51	312.03	330.58	350.24	371.06	393.13	416.51	441.27	467.51	495.31	-95
6	262.53	278.14	294.68	312.21	330.77	350.44	371.28	393.36	416.75	441.53	467.78	495.60	94
7	262.69	278.31	294.85	312.39	330.96	350.64	371.49	393.58	416.99	441.78	468.05	495.88	93
8	262.84	278.47	295.02	312.57	331.15	350.85	371.71	393.81	417.23	442.04	468.32	496.17	92
9	262.99	278.63	295.20	312.75	331.35	351.05	371.92	394.04	417.47	442.29	468.59	496.46	91
+10	263.14	278.79	295.37	312.93	331.54	351.25	372.14	394.27	417.71	442.55	468.86	496.74	-90
11	263.29	278.95	295.54	313.11	331.73	351.45	372.35	394.49	417.95	442.80	469.14	497.03	89
12	263.45	279.11	295.71	313.29	331.92	351.66	372.57	394.72	418.19	443.06	469.41	497.32	88
13	263.60	279.27	295.88	313.47	332.11	351.86	372.78	394.95	418.43	443.32	469.68	497.61	87
14	263.75	279.43	296.05	313.65	332.30	352.06	373.00	395.18	418.68	443.57	469.95	497.89	86
+15	263.90	279.59	296.22	313.83	332.50	352.27	373.21	395.41	418.92	443.83	470.22	498.18	-85
16	264.05	279.76	296156	314.02	332.69	352.47	373.43	395.64	419.16	444.09	470.49	498.47	84
17	264.21	279.92	296.56	314.20	332.88	352.67	373.65	395.86	419.40	444.34	470.76	498.76	83
18	264.36	280.08	296.73	314.38	333.07	352.88	373.86	396.09	419.65	444.60	471.04	499.05	82
19	264.51	280.24	296.91	314.56	333.27	353.08	374.08	396.32	419.89	444.86	471.31	499.33	81
+20	264.67	280.40	297.08	314.74	333.46	353.29	374.29	396.55	420.13	445.11	471.58	499.62	-80
21	264.82	280.57	297.25	314.92	333.65	353.49	374.51	396.78	420.37	445.37	471.85	499.91	79
22	264.97	280.73	297.42	315.11	333.84	353.69	374.73	397.01	420.62	445.63	472.13	500.20	78
23	265.12	280.89	297.59	315.29	334.04	353.90	374.94	397.24	420.86	445.88	472.40	500.49	77
24	265.28	281.05	297.76	315.47	334.23	354.10	375.16	397.47	421.10	446.14	472.67	500.78	76
+25	265.43	281.21	297.94	315.65	334.42	354.31	375.38	397.70	421.35	446.40	472.94	501.07	-75
26	265.58	281.38	298.11	315.83	334.62	354.51	375.59	397.93	421.59	446.66	473.22	501.36	74
27	265.74	281.54	298.28	316.02	334.81	354.72	375.81	398.16	421.83	446.92	473.49	501.65	73
28	265.89	281.70	298.45	316.20	335.00	354.92	376.03	398.39	422.08	447.17	473.76	501.94	72
29	266.04	281.86	298.63	316.38	335.13	355.13	376.24	398.62	422.32	447.43	474.04	502.23	71
+30	266.20	282.03	298.80	316.57	335.39	355.33	376.46	398.85	422.56	447.69	474.31	502.52	-70
31	266.35	282.19	298.97	316.75	335.58	355.54	376.68	399.08	422.81	447.95	474.59	502.81	69
32	266.51	282.35	299.14	316.93	335.78	355.74	376.90	399.31	423.05	448.21	474.86	503.10	68
33	266.66	282.52	299.32	317.11	335.97	355.95	377.11	399.54	423.30	448.47	475.13	503.39	67
34	266.81	282.68	299.49	317.30	336.17	356.15	377.33	399.77	423.54	448.73	475.41	503.68	66
+35	266.97	282.84	299.66	317.48	336.36	356.36	377.55	400.00	423.79	448.99	475.68	503.97	-65
36	267.12	283.01	299.84	317.66	336.55	356.57	377.77	400.23	424.03	449.25	475.96	504.26	64
37	267.28	283.17	300.01	317.85	336.75	356.77	377.99	400.46	424.28	449.50	476.23	504.55	63
38	267.43	283.33	300.18	318.03	336.94	356.98	378.21	400.69	424.52	449.76	476.51	504.84	62
39	267.59	283.50	300.36	318.22	337.14	357.18	378.42	400.93	424.77	450.02	476.78	505.14	61
+40	267.74	283.66	300.53	318.40	337.33	357.39	378.64	401.16	425.01	450.28	477.06	505.43	-60
41	267.90	283.83	300.70	318.58	337.53	357.60	378.86	401.39	425.26	450.54	477.34	505.72	59
42	268.05	283.99	300.88	318.77	337.72	357.80	379.08	401.62	425.50	450.81	477.61	506.01	58
43	268.21	284.15	301.05	318.95	337.92	358.01	379.30	401.85	425.75	451.07	477.89	506.30	57
44	268.36	284.32	301.22	319.14	338.11	358.22	379.52	402.09	426.00	451.33	478.16	506.60	56
+45	268.52	284.48	301.40	319.32	338.31	358.42	379.74	402.32	426.24	451.59	478.44	506.89	-55
46	268.67	284.65	301.57	319.50	338.50	358.63	379.96	402.55	426.49	451.85	478.72	507.18	54
47	268.83	284.81	301.75	319.69	338.70	358.84	380.18	402.78	426.73	452.11	478.99	507.48	53
48	268.98	284.98	301.92	319.87	338.89	359.05	380.40	403.02	426.98	452.37	479.27	504.44	52
49	269.14	285.14	302.10	320.06	339.09	359.25	380.62	403.25	427.23	452.63	479.55	508.06	51
+50	269.29	285.30	302.27	320.24	339.29	359.46	380.84	403.48	427.47	452.89	479.82	508.36	-50
51	269.45	285.47	302.44	320.43	339.48	359.67	381.06	403.71	427.72	453.15	480.10	508.65	49
52	269.60	285.63	302.62	320.61	339.68	359.88	381.28	403.95	427.97	453.42	480.38	508.94	48
53	269.76	285.80	302.79	320.80	339.87	360.08	381.50	404.18	428.22	453.68	480.66	509.24	47
54	269.91	285.96	302.97	320.98	340.07	360.29	381.72	404.42	428.46	453.94	480.93	509.53	46
+55	270.07	286.13	303.14	321.17	340.27	360.50	381.94	404.65	428.71	454.20	481.21	509.83	-45
56	270.23	286.30	303.32	321.36	340.46	360.71	382.16	404.88	428.96	454.47	481.49	510.12	44
57	270.38	286.46	303.49	321.54	340.66	360.92	382.38	405.12	429.21	454.73	481.77	510.41	43
58	270.54	286.63	303.67	321.73	340.86	361.13	382.60	405.35	429.45	454.99	482.05	510.71	42
59	270.70	286.79	303.85	321.91	341.05	361.33	382.82	405.58	429.70	455.25	482.32	511.00	41
+60	270.85	286.96	304.02	322.10	341.25	361.54	383.04	405.82	429.95	455.52	482.60	511.30	-40
61	271.01	287.12	304.20	322.28	341.45	361.75	383.26	406.05	430.20	455.78	482.88	511.60	39
62	271.16	287.29	304.37	322.47	341.65	361.96	383.49	406.29	430.45	456.04	483.16	511.89	38
63	271.32	287.46	304.55	322.66	341.84	362.17	383.71	406.52	430.70	456.31	483.44	512.19	37
64	271.48	287.62	304.72	322.84	342.04	362.38	383.93	406.76	430.94	456.57	483.72	512.48	36
+65	271.64	287.79	304.90	323.03	342.24	362.59	384.15	406.99	431.19	456.83	484.00	512.78	-35
66	271.79	287.95	305.08	323.22	342.44	362.80	384.37	407.23	431.44	457.10	484.28	513.08	34
67	271.95	288.12	305.25	323.40	342.63	363.01	384.59	407.46	431.69	457.36	484.56	513.37	33
68	272.11	288.29	305.43	323.59	342.83	363.22	384.82	407.70	431.94	457.63	484.84	513.67	32
69	272.26	288.45	305.61	323.78	343.03	363.43	385.04	407.93	432.19	457.89	485.12	513.97	31
+70	272.42	288.62	305.78	323.96	343.23	363.64	385.26	408.17	432.44	458.16	485.40	514.26	-30
71	272.58	288.79	305.96	324.15	343.43	363.85	385.48	408.41	432.69	458.42	485.68	514.56	29
72	272.74	288.95	306.14	324.34	343.63	364.06	385.71	408.64	432.94	458.68	485.96	514.86	28
73	272.89	289.12	306.31	324.53	343.82	364.27	385.93	408.88	433.19	458.95	486.24	515.15	27
74	273.05	289.29	306.49	324.71	344.02	364.48	386.15	409.11	433.44	459.22	486.52	515.45	26
+75	273.21	289.45	306.67	324.90	344.22	364.69	386.38	409.35	433.69	459.48	486.80	515.75	-25
76	273.37	289.62	306.84	325.09	344.42	364.90	386.60	409.59	433.94	459.75	487.08	516.05	24
77	273.52	289.79	307.02	325.28	344.62	365.11	386.82	409.82	434.19	460.01	487.37	516.35	23
78	273.68	289.96	307.20	325.47	344.82	365.32	387.05	410.06	434.44	460.28	487.65	516.64	22
79	273.84	290.12	307.38	325.65	345.02	365.53	387.27	410.30	434.70	460.54	487.93	516.94	21
+80	274.00	290.29	307.55	325.84	345.22	365.74	387.49	410.53	434.95	460.81	488.21	517.24	-20
81	274.16	290.46	307.73	326.03	345.42	365.96	387.72	410.77	435.20	461.08	488.49	517.54	19
82	274.32	290.63	307.91	326.22	345.62	366.17	387.94	411.01	435.45	461.34	488.77	517.84	18
83	274.47	290.80	308.09	326.41	345.82	366.38	388.17	411.25	435.70	461.61	489.06	518.14	17
84	274.63	290.96	308.26	326.60	346.02	366.59	388.39	411.48	435.95	461.88	489.34	518.44	16
+85	274.79	291.13	308.44	326.78	346.22	366.80	388.61	411.72	435.20	462.14	489.62	518.74	-15
86	274.95	291.30	308.62	326.97	346.42	367.01	388.84	411.96	436.46	462.41	489.91	519.04	14
87	275.11	291.47	308.80	327.16	346.62	367.23	389.06	412.20	436.71	462.68	490.19	519.34	13
88	275.27	291.64	308.98	327.35	346.82	367.44	389.29	412.44	436.96	462.94	490.47	519.64	12
89	275.43	291.80	309.16	327.54	347.02	367.65	389.51	412.67	437.21	463.21	490.76	519.94	11
+90	275.59	291.97	309.34	327.73	347.22	367.86	389.74	412.91	437.47	463.48	491.04	520.24	-10
91	275.75	292.14	309.51	327.92	347.42	368.08	389.96	413.15	437.72	463.75	491.32	520.54	9
92	275.90	292.31	309.69	328.11	347.62	368.29	390.19	413.39	437.97	464.01	491.61	520.84	8
93	276.06	292.48	309.87	328.30	346.82	368.50	390.41	413.63	438.22	464.28	491.89	521.14	7
94	276.22	292.65	310.05	328.49	348.02	368.71	390.64	413.87	438.48	464.55	492.17	521.44	6
+95	276.38	292.82	310.23	328.68	348.22	368.93	390.86	414.11	438.73	464.82	492.46	521.74	-5
96	276.54	292.99	310.41	328.87	348.42	369.14	391.09	414.35	438.98	465.09	492.74	522.04	4
97	276.70	293.16	310.59	329.06	348.62	369.35	391.32	414.59	439.24	465.36	493.03	522.35	3
98	276.86	293.33	310.77	329.25	348.83	369.57	391.54	414.83	439.49	465.63	493.31	522.65	2
99	277.02	293.50	310.95	329.44	349.03	369.78	291.77	415.06	439.75	465.89	493.60	522.95	1
+100	277.18	293.66	311.13	329.63	349.23	369.99	392.00	415.30	440.00	466.16	493.88	523.25	0
	$C^\sharp_4$	D_4	$D^\sharp_4$	E_4	F_4	$F^\sharp_4$	G_4	$G^\sharp_4$	A_4	$A^\sharp_4$	B_4	C_5	CENTS FLAT

c. The deviations in cents of intervals for a pipe organ tuned in the Werckmeister III keyboard temperament for a tonic of G_4 are major second, -10; major third, +11; major fourth, +5; major fifth, -6; major sixth, +10; and major seventh, +3 (from Table 16.1, p. 309 of Fundamentals of Musical Acoustics). List the frequencies for these notes. (Note: Werckmeister's procedure for a harpsichord would not result in these frequencies.)

16-4. Consider the list of frequencies for the partials of a piano string as found on p. 315 of Fundamentals of Musical Acoustics by A. H. Benade. As an exercise only in locating the named notes and deviations for various repetition frequencies, find this designation of the components 261.63, 523.51, 785.91, 1049.23, 1313.23, and 1578.68 Hz. Before you form any possibly misleading musical conclusions, work problem 16-5 below.

16-5. Consider an idealized vibrating string as described in numerous physics and mathematics textbooks on vibration. This string is imagined to vibrate with modes that are exact integer multiples of mode 1. You are to consider the "problem of the dissonant seventh partial."

a. Assume that the repetition frequency of mode 1 is 261.63 Hz (C_4) and calculate the frequency of mode 7.

b. Find the named note in equal temperament and deviation of the repetition rate found in part a for mode 7.

c. The beat rate between mode 7 and the closest named note in equal temperament is rather large. Verify that this does not mean that mode 7 of this string would form dissonant beats with any mode of another string with similar properties arranged so that mode 1 of the second string and mode 1 of the first string formed a beatless perfect fifth. (See Table 16.1 above.) (Hint: You may wish to refer back to the use of heterodyne frequencies as discussed in chapter 14 of Fundamentals of Musical Acoustics by A. H. Benade. See also problem 14-4.)

16-6. It is traditional to use the "just major scale" to discuss the problems of key modulation. Start with the equal temperament scale C_4 and find the frequency of a major second above it. Now use the same sequence of intervals you used in problem 16-2b above and this newly determined D_4 as the tonic to obtain the list of frequencies for a D major "just scale." Comment on the frequency differences you would find if D major "just scale" intervals were desired from a keyboard instrument tuned to a C major "just scale."

16-7. An interesting comparison can be made between the "just scale" and that of equal temperament. Recall that modulation from one key to another in the "just scale" presents some problems that arise from the wish to have the exact ratios of frequencies listed in Table 16-2 above. Let us consider the degree of deviation of the scale of equal temperament and the "just scale." Since we deal with ratios of frequencies relative to the starting note, the same relationships will exist for the comparison of the "just scale" to the equal temperament scale for each semitone of the equal temperament keyboard.

The relationship for a single note as starting point can be seen from a plot of the deviations in cents between the two scales. A suggested form for the plot and one entry is shown in the accompanying diagram. Plot the deviation in cents of the just ratios as compared to the 100 cent semitone of equal temperament scale. The deviation of the minor second is 200 - 182.404 = +17.596 cents which is approximately 17.6 cents.

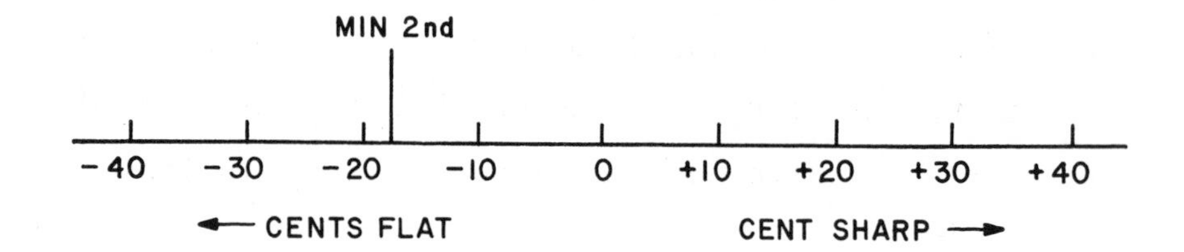

DEVIATION BETWEEN JUST AND EQUAL TEMPERAMENT FOR AN ARBITRARY TONIC

a. First, plot the portion of the diagram that is formed from only the major and minor second, third, fourth, fifth, sixth, and seventh intervals. They should fall into three clusters: lower (flat), central, and upper (sharp). What is the width or range for each cluster specified in cents? What is the approximate middle value in cents of each cluster?

b. Second, plot the deviations of the rest of the intervals from Table 16.2. Do all of them fall within the clusters found in part a above? List those that fall outside these clusters. Did the grouping into three clusters remain when all the deviations were entered in the diagram?

c. It is possible to explain how a performer who wished to play the instrument in just intonation with a major or minor "just scale" could develop the habit of thinking sharp or flat compared to the equal temperament keyboard tuning. Is there a special grouping that separates the major and minor "just scales"?

16-8. The fretted string instruments represent a group of fixed pitch (or very nearly so) instruments. They share many of the temperament and tuning considerations found in keyboard instruments. In his book on Tuning and Temperament-a Historical Survey, J. Murray Barbour observes that Vincenzo Galilei (Galileo's father) was probably the first to use the "rule of eighteen" in placing frets along the string of a lute. He used the ratio 18/17 and observed that others such as 17/16 and 19/18 would not serve the purpose. A very old mathematical instrument, the Mesolabium, was used to find mean proportionals mechanically. An illustration of this device is to be found on page 51 of Barbour's book. When the "rule of eighteen" is combined with the ideas of the Mesolabium, a compass and ruler method can be devised to lay out the fret position. A diagram such as those found in books on instrument construction is given below. The distance L is the length of the string between the bridge and the proposed nut position. A line perpendicular to the string is drawn at the location of the nut. The length of this line is L/18. A hypotenuse is drawn to complete the triangle. The position of the nut is the center of a circular arc with radius L/18 that locates the first fret. At the position of the first fret draw a perpendicular line. Then use the distance from the string to the hypotenuse at the first fret to draw the arc of a circle that will locate the second fret. This procedure is continued until the twelfth fret is located.

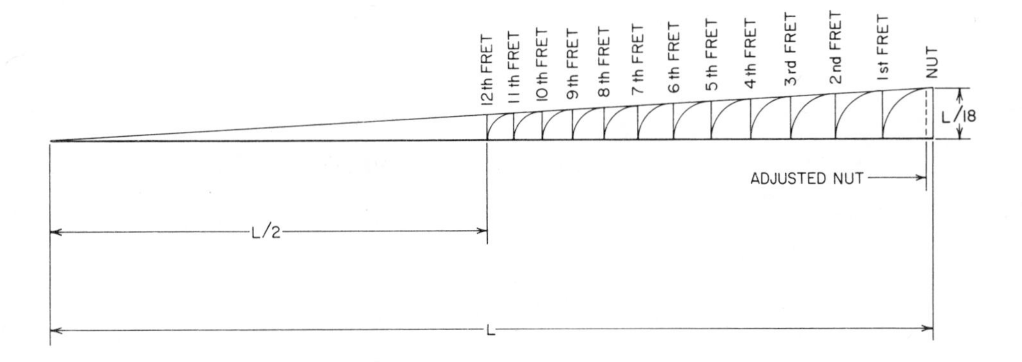

a. Compare the ratio of the second or semitone in this scheme to that of equal temperament. In addition to calculating the ratio accurate to six decimal places, find the interval in cents. Compare the ratio used in the rule of eighteen to those said not to work—17/16 and 19/18.

b. Vincenzo stated that the twelfth fret would be at the center of the string. If this is so, does one have to allow for the width of the line drawn by the compass and a bit of uncertainty in placing the center of the compass when setting the octave? A careful builder will attend to such things and place the octave as well as the frets in their proper positions. (I would expect any adjustments would be consistent with the proportions in the "rule of eighteen"—as experience dictates.)

c. Barbour reports that "Mersenne testified that Galilei's method was favored by 'many makers of instruments'." Consider this observation in regard to one other adjustment made for fretted instruments. When the string is pressed against the fret the string is lengthened and the tension increased slightly. To compensate for the fact that the open string would not be "in tune" with the stopped strings, the nut alone is repositioned slightly closer to the bridge. Discuss the effects of the repositioning of the nut and of stopping the string at a fret with the fingertip in relation to section 16.5 of Fundamentals of Musical Acoustics by A. H. Benade. The adjustment will bring the open string into agreement with the octave. Can one give a firm rule for the change, or will it depend on the material used for the string?

d. Before finishing the discussion of fretted string instruments, obtain measurements of the fret positions from an actual musical instrument. Calculate and report the frequency ratios that result from this fret spacing. Comment on the technique of "pushing" or "pulling" the string with the finger at the fret so as to affect the pitch of the sound or frequency of the modes. Finally, just how close was Vincenzo's method to modern equal temperament?

16-9. It is all too easy to become enraptured by the numerical relationships of various scales that one can devise. The just minor scale given by A. H. Benade in Fundamentals of Musical Acoustics, page 305, has the minor second with a ratio of 10/19. Find the ratios of adjacent notes for the ratios given there. Now consider these same ratios if the second in the just minor scale is a major second with a ratio of 9/8. Extend the discussion to include the other intervals along the lines of section 15.3. To help in visualizing this in terms of the deviation between "just" and "equal temperament" refer to problem 16-5. Comment on the rules or symmetries that favor one interval for the second or the other. Reserve judgment of which is "right" until you finish the rest of the questions in this chapter.

16-10. Problem 15-4 describes a performance task for a woodwind trio. In that problem the intervals were to be performed as perfect minor sixths and major thirds. Use the tables for equal temperament in this chapter and answer the same questions.

a. How would this composition sound in equal temperament compared to the just or perfect intervals of problem 15-4?

b. How would this sound on a vibrato-free electric or pipe organ? Comment on the steps to take if this passage were used in keyboard (organ) and woodwind music.

16-11. An essential point of including certain musical passages in a discussion of scales is to emphasize the fact that intonation is a real part of musical performance. It is a genuine part that requires attention at all times, but there is no set of pat answers. When something is wrong, however, the messages are usually clear. Consider the following pairs of notes to be played by skilled players with no distractions of musical context.

a. Assume C_4 has a repetition rate of 261 Hz and calculate the frequency associated with $G_4^{\#}$ from Table 16-2. Then use this value of the repetition rate for $G_4^{\#}$ to calculate the frequency of C_5 implied by this $G_4^{\#}$.

b. Use the data of part a to describe what careful musicians would do while performing the following malicious passage.

16-12. Use the data in question 16-10 to describe what three woodwind players might do in the performance of this passage.

16-13. Consult <u>Fundamentals of Musical Acoustics</u> by A. H. Benade, section 15.3 and Figures 15.2 and 15.3, as you consider these passages. Find and comment on the traps included.

It would be a very good idea to try these, and those in 16-11 and 16-12, for yourself. Any combination of flute, sax, trumpet, or oboe—which should be played a fifth higher on the scale of the instrument to keep the task practical—could be played. It won't work for bowed strings or keyboard instruments.

16-14. It is axiomatic that bugles play loudly. Discuss the musical implications of a rather simple duet based on a military quickstep. For simplicity assume that the bugles are generating appreciable amounts of the first four harmonics as is characteristic of British bugles. Next,

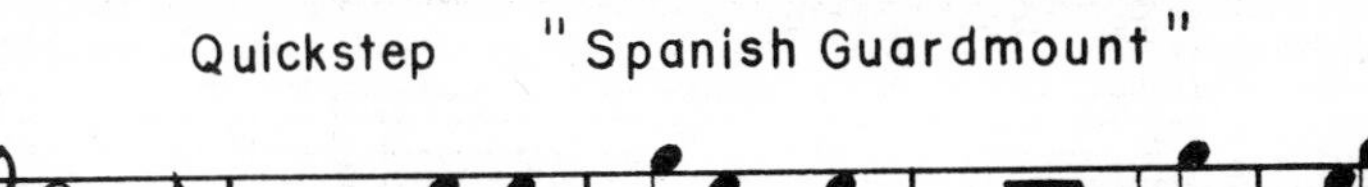

discuss the musical implications if the listener is far enough away from the sound that no heterodyne frequencies are generated. What is the pedal note of the bugle? If you wish to simplify the calculations you can assume the bugle plays the written C at 250 Hz.

16-15. The human voice and bowed string instruments were very important in shaping the structure of pre-Renaissance music. These instruments produce tones with a loudness spectrum that is approximately the same. The spectra contain a considerable number of harmonics with appreciable strength whose amplitudes become progressively smaller with increasing frequency. Let us treat these sounds as if they were sufficiently alike to be equivalent. In church music, the sound pressure at the location of the listener tends to be relatively low. Write a brief discussion to show the basic acoustical reasons for some of the intervals and tuning methods used in this early music.

Some knowledge of the aspects of pre-Renaissance music will be helpful. Vibrato was not used. The music required a strict beat-free intonation. The intervals of unisons, thirds, fifths, and octaves along with simple chords of the major and minor keys were about all that were used by composers. Recall that beats up to 20/sec are acceptable and frequency differences beyond that tend merely toward roughness. The notes used in music are generally in the range A_2 to A_5. A preponderance of "unorganized" sound will mask, or obscure, the noticeability of the internal organization of parts of the sound reaching our ears.

17

Sound Production in Pianos

17-1. Let us begin this question with a set of assertions. At a place near B_2 on a piano there is a change between triply-strung notes belonging to the upper ranges of the keyboard, and the single over-wound heavy strings of the lower notes. The wave impedance of each string above the break (immediately above, that is) turns out to be about 70 percent as large as the wave impedance of the wound string directly below the break.

a. Taken at face value this might imply that the decay time of the unwound triply-strung notes would be longer than the decay time of the wound-string notes just below them at the break. Explain this remark.

b. Examine a real piano in the neighborhood of the break and verify that on a decent instrument the decay or halving times on each side are in fact rather similar. Discuss the way this equalization is accomplished.

c. Explain why it is proper in these considerations to make use of the wave impedance of a single member of the triply-strung set rather than the three-fold larger wave impedance that we would calculate for the three strings taken as a single entity.

17-2. If one holds one end of a light wooden stick and bounces it off of the strings of the note C_4 of a piano (the damper being lifted for a short time) one observes that the following string sets will respond sympathetically as a forced oscillation or resonance:

C_5 G_5 C_6 G_6 C_7 G_7 ... not much beyond this ...

 If, on the other hand, when the string of note C_4 is struck by a certain rubber mallet the same experiment fails to show excitation of sympathetic strings beyond G_6. Describe how you would obtain a numerical value for the hammer contact time in this case. Give the number you get from such a calculation, and show your calculations.

17-3. Suppose a piano is beautifully proportioned to play when tuned with a scale based on A-440 Hz. List some changes that you would recommend to its maker in his choice of string length and thickness, hammer weight, soundboard stiffness, and the like if he is setting out to design an equally good piano to play at A-420 Hz. If you prefer, simplify the discussion by assuming that the string lengths are to be maintained equally on both instruments. Recall that we are very interested in the effects of inharmonicity (of various kinds), in the duration of the various notes, in the nature of the vibrational recipe (as influenced by hammer contact time, etc.), and in the feel of the hammers via the keys at the player's fingers.

17-4. It is often said that the seventh member of a harmonic series is dissonant. Supposedly large efforts are made to remove it from the tone of a piano or other instrument, while in fact this component cannot be removed from piano tones and is normally present in all other musical tones as well, often in considerable amounts, with no ill effects. Consider a tone made up of a large number, say 15, of the harmonics of 100 Hz, and another tone Y whose components are many of the multiples of 700 Hz. Identify enough heterodyne components to decide whether there will be any beats or other signs of roughness. Consider also the possibilities for tones based on 400 Hz or 500 Hz, along with the 700 Hz tone that is considered troublesome occasionally.

See if, on the other hand, plausible reasons for the growth of this traditional belief in the dissonance of the 7th harmonic can be stated. After all, very few musicians can be fooled on such a point, so one should be able to find reasons for the belief.

17-5. A very traditional method for showing musical relationships is by means of a "monochord." Ptolemy (second century), in his directions for constructing the monochord, stated that a string of consistent diameter be stretched between two bridges of equal height and that a third movable bridge, somewhat higher than the end bridges, be provided to divide the string. In a modern physics-demonstration version the string passes over a pulley to a weight which keeps it under tension. Historically this instrument has been used as a teaching device. It is customary to demonstrate the various musical intervals (unison, octave, fifth, third, etc.) by plucking the string with the interbridge distance reduced by 1/2, 2/3, 4/5, etc. in accordance with the simple theory of strings that leads to harmonics. If the two segments of a string divided by a bridge are plucked one is safe in assuming the tension is the same in each segment. Their frequencies should depend inversely upon the length of the segment. If two monochords are used to demonstrate the intervals, great care must be made to use strings of the same diameter, composition, and tension for each monochord.

a. Careful-listening experiments invariably show significant discrepancies between "theoretical expectation" and what actually happens. List some of the properties of real strings that would give rise to these discrepancies.

b. Consult your list and decide the relative magnitudes of the pitch discrepancies expected for the length ratios for the octave (1:2), fifth (2:3), and third (4:5).

 In order to compare the results of your analysis for these intervals, use the pitch of the original full-length string as the reference. Are the discrepancies on the sharp or flat side of the theoretical interval?

c. Discuss the reasons behind the observations:

1. A gut string tends to have somewhat greater discrepancies than does a nylon string of the same size.
2. Both of them show larger discrepancies than a steel string sized so it sounds the same pitch at the same full length.

17-6. Imagine that you are holding down the C_4 key on a piano, so that this set of strings is left undamped, but not excited by its own hammer. Now imagine that the G_4 key is given a short sharp blow and released (so that the G_4 strings are vigorously excited and then abruptly damped into silence). Nevertheless the piano continues to emit a tone via the undamped C-strings. Explain what is happening in terms of the ways the strings interact with the soundboard and the soundboard with the strings. What is the pitch of the sound?

17-7. The partials belonging to an actual (but slightly simplified) piano string at the note C_4 will, because of the effect of stiffness, have frequencies such as the following:

261.63, 523.51, 785.91, 1049.23, 1313.23, 1578.68 Hz

Notice that these partials are not quite harmonic in their frequency ratios. A piano tuner will normally set the tuning of the octave C_5 for minimum jangle which will lead him to provide the upper string with the following collection of partials:

523.70, 1048.81, 1571.11 Hz

a. Work out the clumps of components that form for the lower note C_4 taken alone as a result of heterodyne action in our ears. Comment briefly on the musical implications of these components.

b. Next work out the additional heterodyne components near 261 Hz that arise from the interaction of C_5 with C_4. (Hint: There are only a few pairings from the two tones that give components in their neighborhood, so don't spend a lot of time searching for a long list of them.) Write a sentence or two more about the musical implications you recognize to be contained in the use of piano-type sounds whose partials are very slightly inharmonic.

18

The Clavichord and the Harpsichord

18-1. In the author's laboratory the properties of a Zuckerman harpsichord, a "Flemish VI," are being studied. With our spectrum analyzer we found the components of the sound of the harpsichord for A_3, as sketched below.

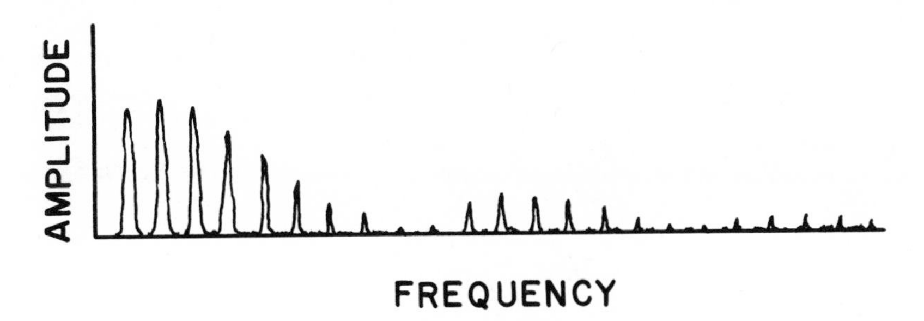

The details of recording this sound via microphone placement need not concern us. It was about 1 meter above the instrument in a room with lots of absorption and no pronounced room modes. The amplitudes are those of the components as they appeared about 150 msec after the initial sound. The graph below represents the appearance of the amplitudes of the components 5 sec after the initial sound.

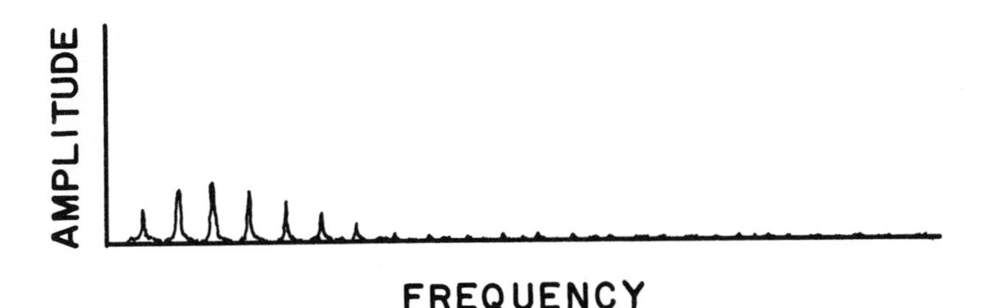

The sound showed inharmonicity in the measurements of the frequency of each partial. The values of frequency in Hz for the first ten partials are given below.

Component	Frequency	Component	Frequency
1	220.0	6	1326.5
2	440.2	7	1544.8
3	660.8	8	1773.2
4	882.2	9	1997.8
5	1104.2	10	2222.9

a. From the distribution of amplitudes among the partials roughly determine where the plectrum is located along this string.

b. Find the amount of change and comment on the shift in pitch with time that would result from a comparison of the two sets of data.

18-2. In many harpsichords one finds two sets of strings, the 4 ft and 8 ft strings. The diagram schematically indicates the arrangement of the hitch-pin rail, bridges, and string terminations in a conventional harpsichord.

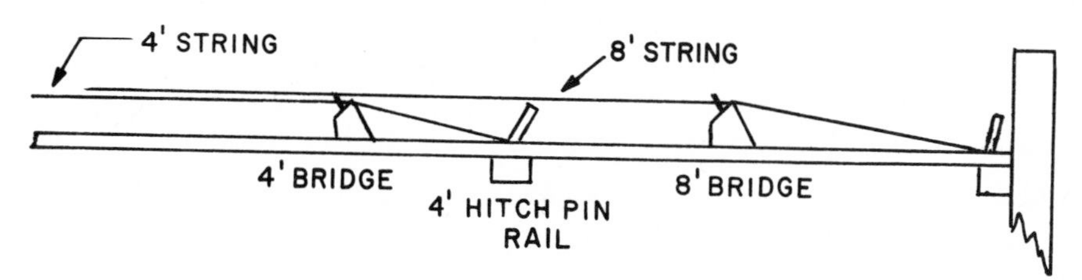

Suppose that you wished to build an experimental harpsichord in which all the strings are attached to a hitch-pin rail on the soundboard as is normally done for the 4 ft strings. The proposed scheme would result in the addition of an 8 ft hitch-pin rail as shown below.

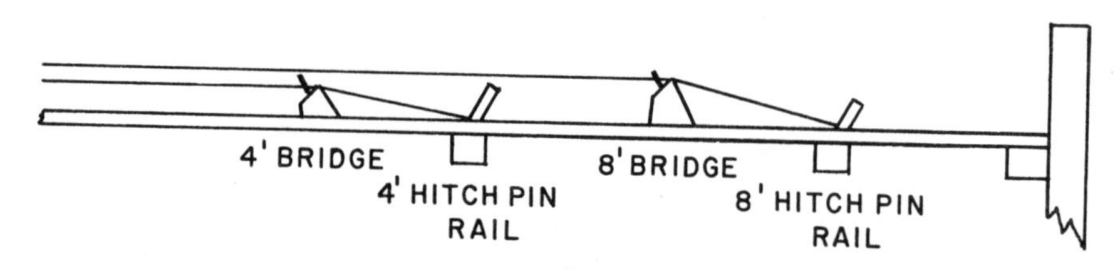

Before building such an instrument you should consider the consequences of transferring the tension of the string to a hitch pin located on the soundboard compared to one at the edge of the soundboard and the bent side. Assume that this change will affect only the behavior of the 8 ft strings.

a. What changes would you expect in the wave impedance relationships between each string and the soundboard?

b. How would you expect this nonstandard method of terminating the string to alter the overall decay times of the various partials in the string sound?

c. Estimate the effect of this nonstandard method of terminating the string on the inharmonicity of the partials.

d. If one insisted on terminating the strings in this manner, how would you go about reproportioning the strings to eliminate or reduce the changes?

18-3. An experimental harpsichord was built not too long ago in which the entire string tension was passed on to the soundboard by a rib that served as a bridge-like structure and hitch rail simultaneously. The soundboard was constructed with an offset at this rail so that no twisting resulted from the tension. A diagram of this arrangement may assist your visualization of the situation.

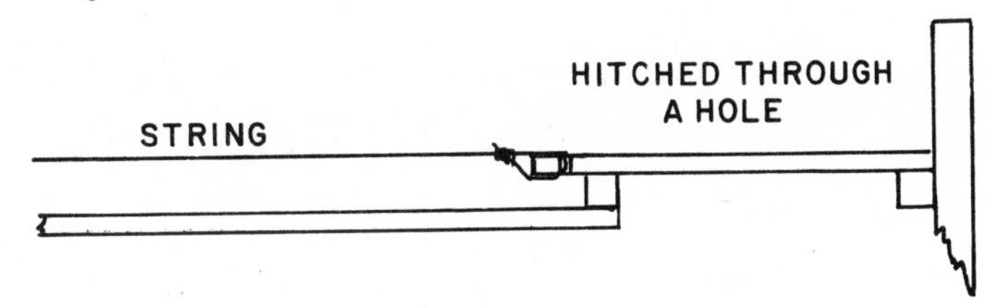

Care was taken to keep all the elastic, damping, etc. parameters of the new structure as similar to the conventional one as possible. When the same materials are used in the conventional structure they work reasonably but not excellently well, it is reported. The experimental structure placed the soundboard between the bridge and side under tension. Consider the effects of such changes and discuss the related aspects of the behavior of the strings.

a. What changes would you expect in the wave impedance relationships between each string and the soundboard in the new arrangement were it placed in longitudinal tension?

b. How would the wave impedance relationships alter the overall decay times of the various partials in a string sound and the degree of their inharmonicity? Comment on the possible changes in the pitch with time.

c. Suggest a way of reproportioning the strings to eliminate or reduce the changes in the sound and produce a more usual one.

18-4. Despite the fact that corrections were made to solve the problems raised in problem 18-3, the experimental harpsichord proved to have an unsatisfactory tone. The vibrational behavior of a board, in contrast to that of a membrane under tension, can form the reason for the underlying physics reason for the particular tone. Presume that the behavior of a board and a membrane can be used to explain the problem with the tone reported for the experimental instrument. The average spacing of the vibrational mode frequencies of a plate (or harpsichord soundboard) is constant over the entire frequency range. For a membrane under tension (a drum head) the spacing is inversely proportional to frequency.

a. Assume that the experimental method of fixing the ends of the strings to holes in the edge of the soundboard placed it under more tension and less compression than otherwise. Approximate the variation in the intermode frequency spacing for a soundboard under one-dimensional tension. Explain your reasoning and support your contention of increasing or decreasing intermode spacing.

b. Consult a reference book that diagrams harpsichord construction or arrange to inspect an actual one. Suggest one or more ways for altered bracing or other structural changes that might be applied to this experimental instrument to bring it to more normal behavior. Suppose you are the builder and that you have decided to abandon this experimental instrument. Give your reasons for returning to the traditional methods of construction.

18-5. The buff stop of a harpsichord gently presses a small piece of felt or soft leather against the string near its end. The buff stop touches the vibrating portion of the string at the point where it passes over the nut. This permits a short-lived but distinct sound to exist, whereas a soft damper of roughly the same dimensions gently touching the string near the plucking point serves to quickly silence the string. In particular, deal with the following points in an explanation of the influence of the buff stop on the string vibrations.

a. What are the relative decay times of the various string modes when they are acted on by a damper applied at the pluck point?

b. What are the relative decay times of the same string modes when the buff stop is used to produce its distinctive sound?

c. How are the two sets of decay times described in a and b related to one another?

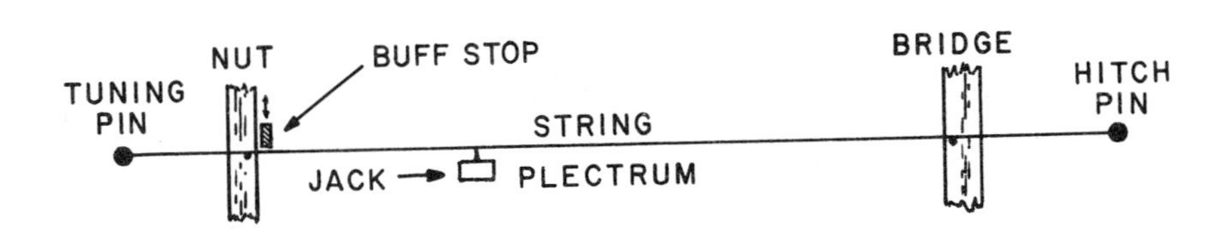

18-6. Consider a hypothetical guitar- or lute-type plucked string instrument. Compare three ways of attaching the strings at the bridge end. In one version the strings pass over the bridge and are attached at the very edge of the soundboard. In a second version the strings pass over the bridge and fasten to the soundboard halfway between the bridge and the edge of the soundboard. In the third version the strings fasten at the bridge.

a. Discuss the differences in wave impedance for the string termination in these three cases. State any simplifying conditions or assumptions you use for your arguments.

b. Consider the top plate modes for a guitar discussed in problem 10-1. Comment on the differences each termination method might have on the various modes.

c. Consult a description and diagram of a clavicord, or inspect an actual one, and identify the bridge and hitch-pin structures. Comment on the possible wave impedance behavior at the bridge, and suggest the general response expected of the soundboard. Evaluate for yourself if the analysis of vibration transfer in problem 10-1 is appropriate for the soundboard of a clavichord.

18-7. The methods of chapter 18 can be applied to the sounds of a banjo-like instrument. The instrument has one of its strings tuned to give the following set of natural vibrational mode frequencies:

Mode	1	2	3	4	5	6	7	8
Frequency	100	199	297	390	510	608	703	801 Hz

a. Make sketches to help you calculate the strengths of the first, fourth, and eighth modes of oscillation when the string is plucked at a distance of one-eighth the distance from the bridge end of the string.

b. Assume that the sound in the room has the component amplitudes similar to the string amplitudes. Given the additional fact that the lower four partials decay more quickly than the upper ones, comment on the initial pitch and the change in pitch with time. (Refer to problem 5-8.)

c. Contrast the pitch change for part b with problem 18-1. How do the heterodyne components and partials compare for each situation?

18-8. Some of the problems presented so far in this chapter have involved sounds with partials that have amplitudes that decrease with increasing frequency. The tone controls of a high-fidelity amplifier can be used to decrease or increase the amplitude of a portion of the audio range. The various possibilities for altering the amplitudes of components of a sound such as in a plucked string can be discussed after a brief description of some of the influences electronic amplifier systems have on amplitude.

The graph below has data that represents a realistic set of high-fidelity tone control characteristics such as you would find in the manufacturer's specifications.

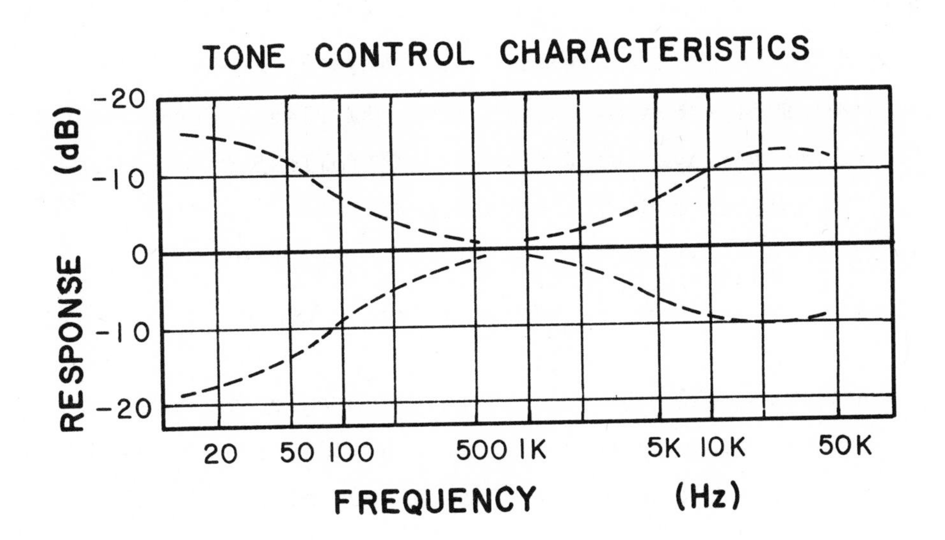

130 The dashed lines represent the extreme range of level adjustment possible for this particular system. Let us consider a situation involving an idealized high-fidelity system where the normal tone control of the amplifier will boost or reduce the signal at a rate of roughly 6 dB per octave on the maximum setting for frequencies beyond a certain "break point frequency" f_b, which is usually near 800 Hz.

In order to discuss the effect of such settings on the amplitude of signals fed into the system consider the following simplifications. The 6 dB per octave specification means that a sound whose frequency components are f_1, f_2, f_3 ... f_n with amplitudes A_1, A_2, A_3 ... A_n will be changed by the treble control to a sound with the same frequencies and new amplitudes a_1, a_2, a_3 ... a_n. The new amplitudes can be calculated for each frequency according to the approximate formula

$$a_n = A_n(f_n/f_b)^m$$

which is to be used only for the components whose frequencies lie above f_b. One uses $m = +1$ when the control is set for full treble boost, $m = +1/2$ when it is set for half boost, and $m = 0$ for flat response. Similarly $m = -1$ corresponds to maximum treble reduction and $m = -1/2$ for an intermediate treble reduction. The bass control operates in exactly the same manner on the frequency components lying below its breakpoint frequency. However the exponent m has the opposite sign from that found for the treble control; that is, $m = -1$ gives the maximum bass boost and $m = +1$ gives maximum reduction.

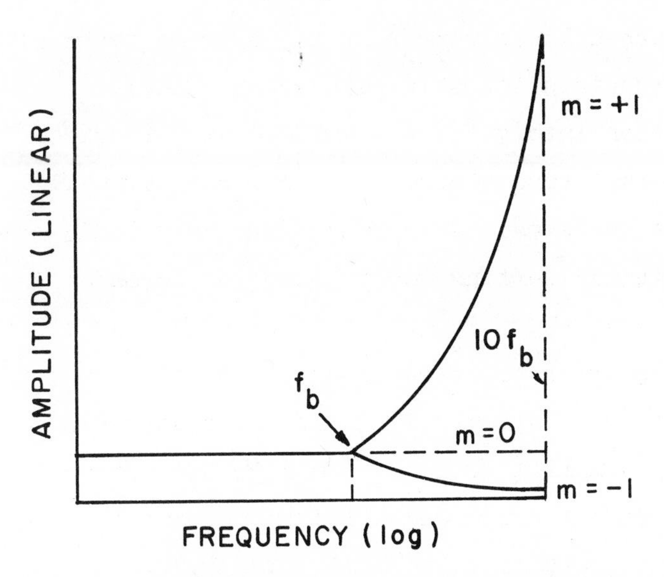

a. Suppose the plucking point and vibration pickup location on a guitar string tuned to 300 Hz are located in such a way that the pickup signal has a recipe of harmonic components whose amplitudes A_n are 10, 5, 3, 2, 1, 0.5, 0.3, 0.1, and 0.05. Calculate the amplitude spectrum of the signal coming from a full-on treble boost tone control if the breakpoint frequency has the usual value of 800 Hz. The bass control is set for a flat response, so the frequencies below the breakpoint are unaffected.

b. Next calculate the amplitude for the same initial sound when the bass control is set at the halfway value $m = +1/2$. The treble control is set for a flat response. Assume that the breakpoint for these calculations is also 800 Hz.

c. Calculate the loudness of the original sound if the loudness of the fundamental component is two sones. Refer back to Chapter 13 if necessary for a review of the effect of the harmonics on loudness. For comparison calculate the loudness of the sound with treble boost discussed in part a.

18-9. In problem 18-8 the effect of the settings of a high-fidelity amplifier tone on the amplitude of the components of a sound are considered. If one sets the bass for maximum reduction and the treble for maximum boost the equation becomes

$$a_n = nA_n$$

for the special case where the sound consists of harmonics or partials that are only slightly inharmonic. Alternatively the setting for maximum bass boost and treble reduction results in the equation

$$a_n = (1/n)A_n$$

a. Verify that both these relationships require the same breakpoint frequency for both controls.

b. It is of interest to consider the effect of the tone control settings in this extreme approximation on a sound that already has a progressive reduction in amplitude with increasing frequency. Consider the effect of both extremes of the control settings on the amplitude of a plucked string sound such as the harpsichord or guitar. Since the objective is to illustrate the effect of the settings rather than a particular sound, you can assume the sound is for a plucked string formed of harmonics with a fundamental of 200 Hz and eight components. List your assumptions for initial amplitude and verify that they are realistic for a plucked string.

19

THE VOICE AS A MUSICAL INSTRUMENT

19-1. A female singer sounds a vowel tone whose fundamental frequency is 500 Hz. Because of her vibrato the first partial of her voice "visits" a small region of her formant around 500 Hz. The higher partials explore the response of the vocal tract in the neighborhood of 1000, 1500 ... Hz. As a result the listener is supplied with enough clues that he can reconstruct the loudness spectrum envelope for the vowel sound of the word "had." Use these clues in the graphical form below to construct a picture of the formant pattern for this vowel.

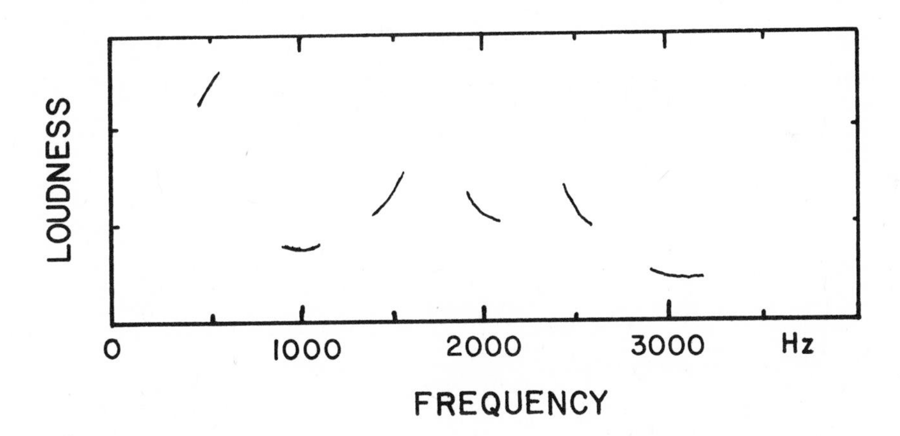

19-2. A singer produces the vowel sound in "heard" without vibrato at a pitch slightly above E_4. The loudness spectrum of her voice partials is indicated below. Utilize the fact that the first formant peak is always the tallest and reconstruct the loudness spectrum envelope for this vowel. (Note: You can locate the possible formant peaks with these data and information in Fundamentals of Musical Acoustics by A. H. Benade, chapter 19.)

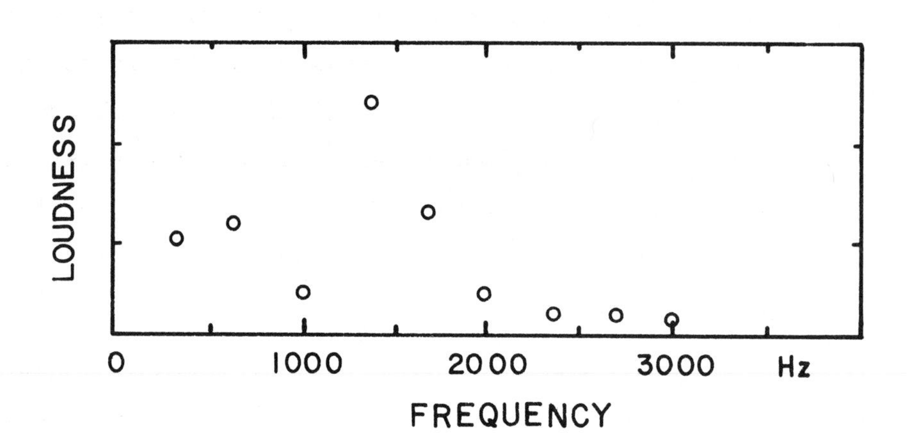

19-3. Consult chapter 19 of Fundamentals of Musical Acoustics by A. H. Benade and an article in the March 1977 Scientific American by J. Sundberg for a discussion of the singing voice. Singers add an extra formant in a process that involves a lowering of the larynx. Assume that the graph below represents a possible loudness recipe for the vowel "ah" sung at A_3. Estimate the relative increase in loudness. Include a discussion of the properties of the listening room.

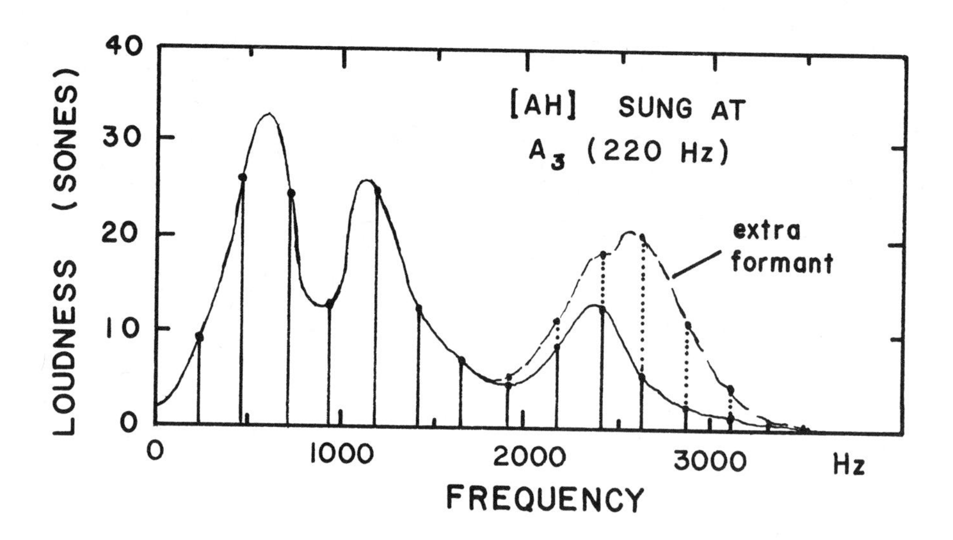

19-4. A singer produces a steady, vibrato-free tone at G_4. The loudnesses of the first five partials are found to be 10, 5, 2, 4, and 2 sones, respectively. When she shifts up a semitone to $G_4^{\#}$ the loudnesses of both fundamental and second harmonic components are decreased. What vowel is she enunciating? Can you comment on the other partials of this vowel being enunciated at $G_4^{\#}$?

19-5. The formant of the vowel "oo" shown below indicates strong peaks at about 300 and 700 Hz. Let us assume that a woman, whose voice formant for this vowel is exactly as shown in the diagram, is asked to sing real music over a pitch range restricted to lie between A_3 and E_5.

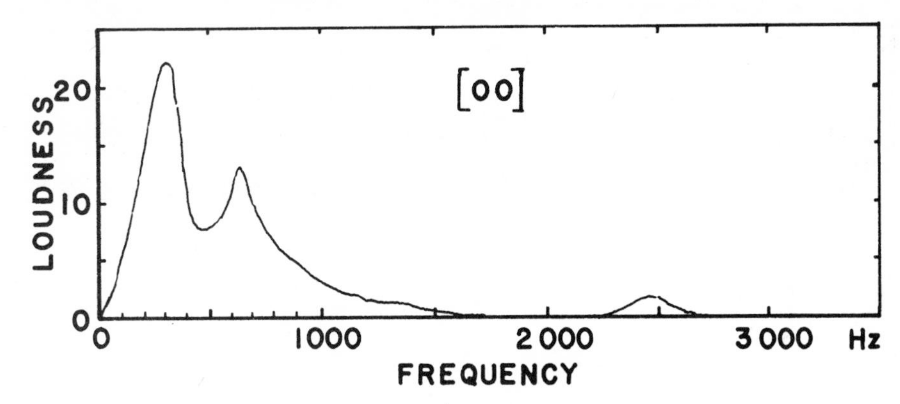

Adapted from Fundamentals of Musical Acoustics, Arthur H. Benade, Oxford University Press, 1976 with permission.

Diagram the loudness recipes for the two extreme pitches listed above and one intermediate pitch. Compare music sung piano-forte (pf) in the key of D major with that sung in the key of F major. Would there be any influence if the music were sung with or without vibrato?

20

THE BRASS WIND INSTRUMENTS

20-1. A trumpet is a little wobbly when played forte on the written G_4. When its player sounds a crescendo from ppp on this note, the pitch first rises a little and then turns wobbly. Tell all you can about the positions of the air column resonance peaks of this instrument. (Hint: Contrast it with the behavior of an instrument whose pitch rises steadily during a crescendo, without getting wobbly.)

20-2. Consider the resonance curve sketched below for a grossly oversimplified sort of brass instrument.

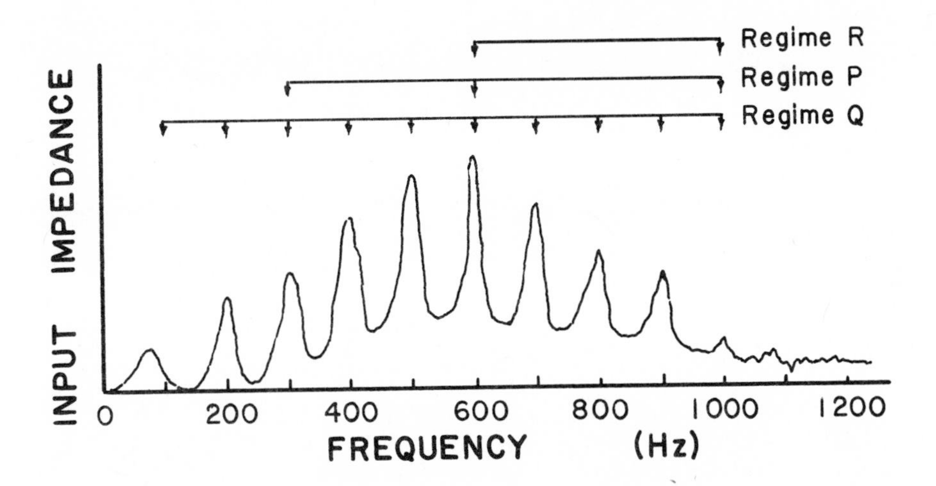

a. Decide whether the regime marked P will run sharp or flat when it is played from pianissimo to fortissimo while the player seeks the best possible response from the instrument. Explain the reasoning behind your decision.

b. A regime of the sort marked Q is possible on a trumpet or trombone. Decide how the tone produced by this sort of regime can include a reasonable amount of the fundamental component in its tone.

c. Explain your reasons for deciding whether a mezzo forte regime of the sort marked P will be easier or harder to play than the one marked R.

d. Assuming that this instrument is a trumpet-like horn for use in an old-time orchestra (A_4 at 461 Hz), figure out the named key of the instrument—that is, does its owner call it a C, B_b, D, A, or what, trumpet?

e. Why would a player in ordinary equal temperament surroundings be prone to complain about the flatness of the regime based on peak 5 (and whatever help comes from peak 10)? If the instrument maker were clever enough to move peak 5 without shifting anything else, so as to cure the alleged problem, would it spoil anything else in the playing behavior of the instrument? Should peak 6 be shifted by the instrument maker?

20-3. Suggest what should be done in trimming the mouthpiece of a trumpet that shows these troublesome symptoms:

a. For horn A the note C_4 is too wobbly to really test, G_4 runs very flat on a crescendo, and C_5 runs flat on a crescendo.

b. For horn B the note C_4 runs slightly sharp on a crescendo, then goes flat; and G_4 runs slightly flat on a crescendo. If you cannot prescribe a cure, at least tell what is wrong with the horn.

c. Which of the notes in the symptoms of horn B will be more wobbly when played at a mezzo forte level?

20-4. It is possible to sound a slightly peculiar tone on the French horn whose (simplified) spectrum of partials is of the following nature.

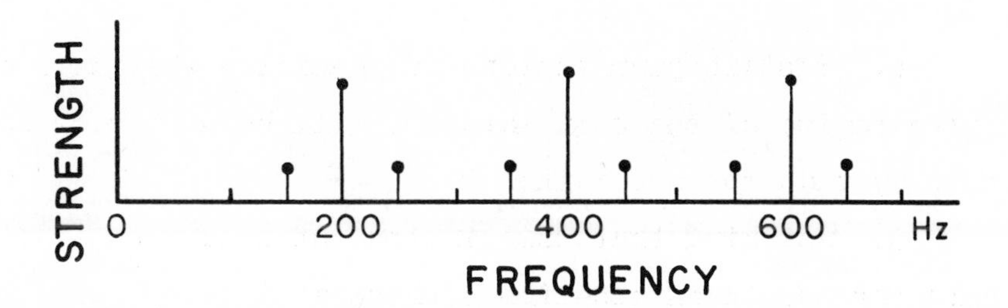

The diagram shows that each harmonic of 200 Hz is supplied with a pair of considerably weaker satellite components spaced 50 Hz above and below it.

a. Estimate the pitch that one might assign to such a tone if it is heard very softly, and then very loudly. Explain your answer.

b. If a duet were to be played between a normal musical instrument and such a horn, one would notice that certain musical intervals would take on an unusual significance. Choose one or two examples of this for explanation, under the assumption that the listener hears the sounds at a low level of loudness.

20-5. Suppose one tries to play as softly as possible with a pressure-controlled reed on a horn whose resonance curve is sketched below.

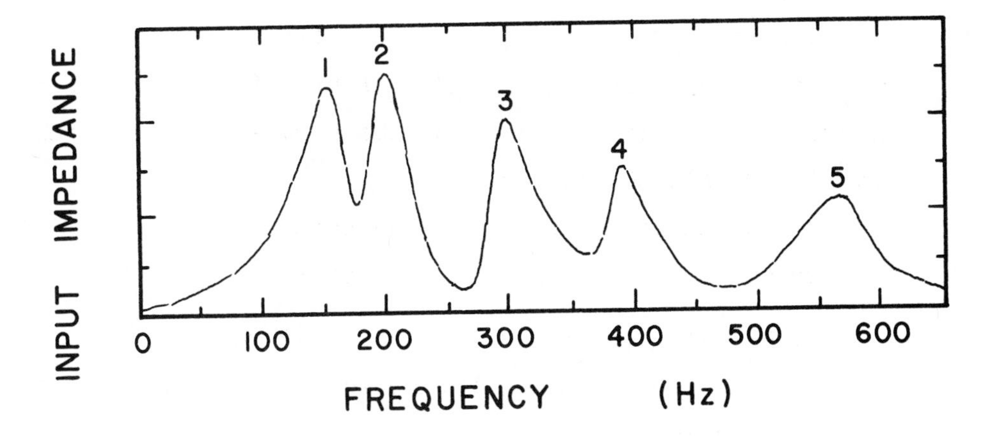

a. Will the sound have a frequency near 150 Hz or near 200 Hz? Explain briefly. Suppose one now plays loudly enough that the first three harmonics of the tone are influential.

 Decide which tone is now favored, and tell which way its pitch will move as the tone is played louder and louder.

b. There is a possible regime of oscillation whose frequency is close to 100 Hz. Use the criteria for finding heterodyne frequencies to show whether this oscillation runs a little above or below 100 Hz.

c. Suppose that while playing a tone which is the one belonging to the 200 Hz regime of oscillation, one pressed a special key whose sole effect is to abruptly displace the 4th resonance peak downward by 50 Hz. What new regime of oscillation would be favored? What would be the second-choice regime?

20-6. Suppose a skilled instrument maker builds a perfectly aligned trumpet in which the resonance peaks 2 through 12 under playing conditions are harmonics located at exact multiples 2, 3 ... 12 of a single frequency. A player who tests the instrument will find it to be responsive and very well tuned. Should he be required to play in the context of equal temperament such as with a piano, he will discover that the note based on peaks 5 and 10 sounds flat.

a. Identify the note name for the regime based on peaks 5 and 10 when you are given that the tone based on the regime 2, 4, 6 ... is written as C_4. Explain how this tone on the trumpet described above is regarded as flat when compared to equal temperament.

b. A skillful craftsman sometimes can use acoustical principles to shift the frequency of a single resonance peak without moving the rest of them. Suppose peak 5 were shifted upward in frequency. Explain why this would not quite solve the problem? What else would the instrument maker have to do to complete the correction? Would these changes spoil anything else about the instrument? If these last changes alter other notes, which notes were altered and by how much were they moved?

21

The Woodwinds

21-1. Consider a saxophone which plays a wide octave (12 cents wide). Explain why increasing the mouthpiece cavity volume enough to lower the pianissimo playing pitch of the low-register tone by 4 cents will come close to fixing the cooperation troubles of the instrument in that part of its scale. Why would the amount of pitch changed by this alteration of the cavity volume be somewhat less for playing experiments at mezzo forte level? (Hint: For present purposes it is reasonable to assume that the number of cooperating peaks on a sax is similar to that for an oboe at corresponding parts of the scale.) Identify the note which you wish to discuss.

21-2. Suppose someone offered you the choice of either of two woodwinds giving only the information contained in the two resonance curves sketched below. Decide what sort of instrument it is, and then make your considered choice on the assumption that it will be used strictly for solo work. Explain your reasoning.

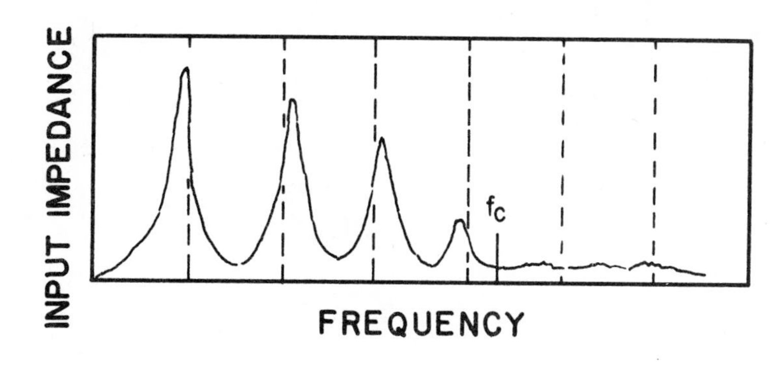

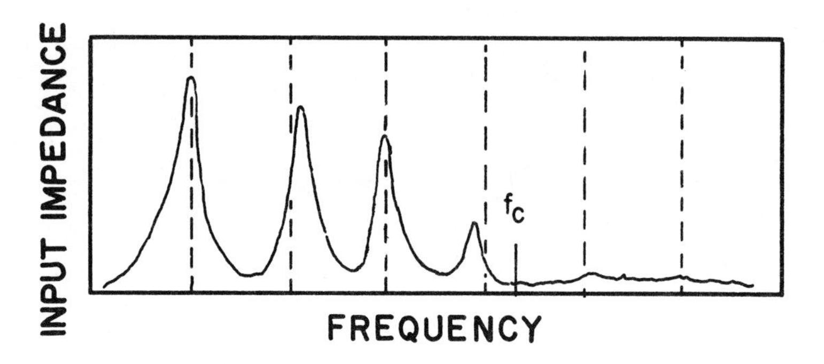

21-3. The register hole of a clarinet makes the peak associated with mode 1 shorter and moves it slightly toward higher frequencies as sketched in the upper resonance curve. Why would it not be a good idea to have things arranged in the manner shown in the lower curve? What would happen if a note were attacked fortissimo under these conditions, and then allowed to diminuendo? Could you describe the resulting tone color? Be prepared for a surprise!

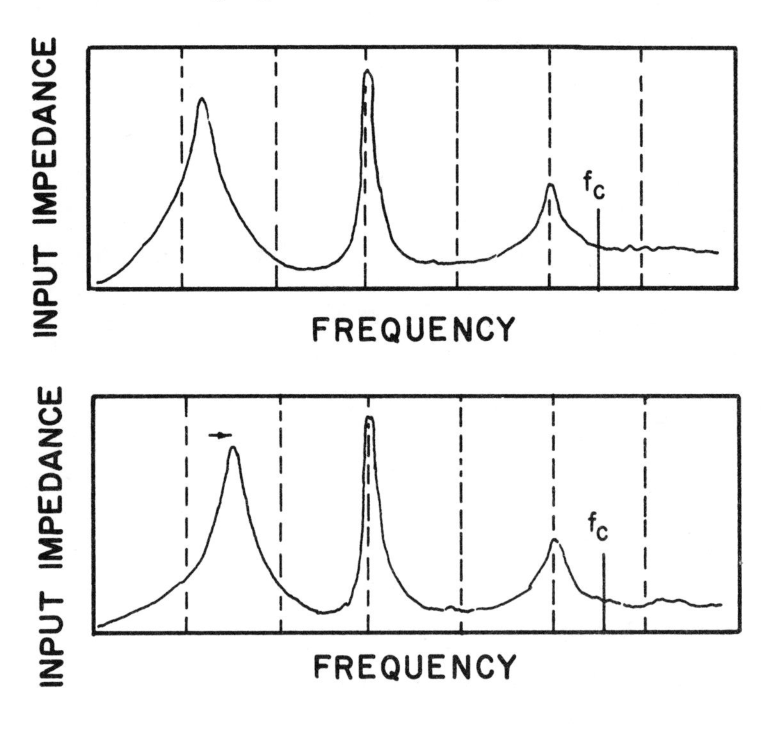

21-4. Other things being equal, a well-behaved and properly singing woodwind has an associated tone color that is closely related to the trend of cutoff frequencies over the fingering scale. Describe and compare the tone color behavior and properties of the two woodwinds having cutoff frequency trends of the sort sketched below.

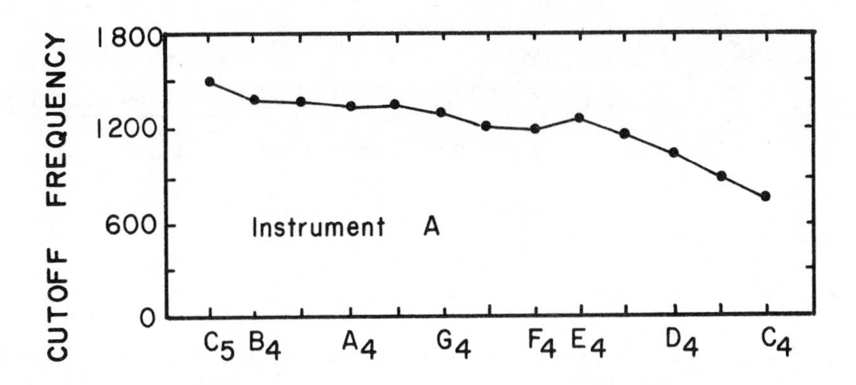

What troubles are implied for oboes for the note B_4?

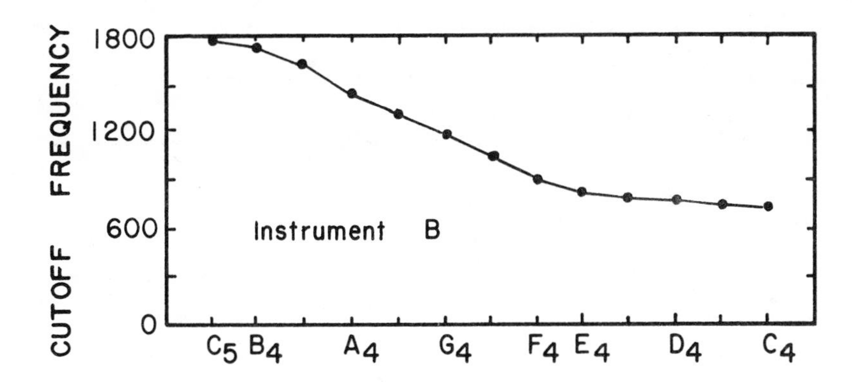

Which instrument would have the easy-to-play and more solid second register?

21-5. The main low-register scale of a C-melody saxophone extends from B_2 to C_4 (although there are keys up as far as E_4 at the upper end that are used to extend only the top part of the second register). As a reasonable simplification one may assume that the open-holes cutoff frequency is about 700 Hz for all the notes in the main scale.

a. Draw the resonance curves to be associated with the fingerings that give the low-register notes C_3, G_3, C_4 for a really good instrument. Be as accurate as possible, giving particular attention to the frequencies and the tallnesses of the various peaks. 1 cm = 100 Hz is a nice frequency scale on the horizontal axis.

b. Now sketch the resonance curves belonging to the second-register notes associated with the octaves to the ones written above. Make the assumption that the register holes provided on the instrument are ideally suited to their purpose. Conclude by stating your judgement of the relative ease of playing these notes and then giving an explanation in a sentence or two.

21-6. Someone has bought an old (1920's) Boehm flute that was properly built to play in tune with a reference pitch of A-435. He measures the instrument and finds that some of the tone holes are located at the positions listed below which are measured from the cork.

At the Top End		In the Middle		At the Bottom End	
C_5	266.5 mm	$G_4^{\#}$	253.6 mm	$C_4^{\#}$	563.9 mm
B_5	285.4 mm	G_4	378.7 mm	C_4	601.5 mm
2s = 19.9 mm		2s = 25.1 mm		2s = 37.6 mm	

(Note: The dimensions labeled 2s are the interhole spacings corresponding to a semitone change in pitch.) In the hopes of playing this flute at the contemporary reference pitch A-440 (which is 20 cents or one-fifth semitone higher than the old one), the owner decides that since a semitone pitch change at G_4 corresponds to a length change in the air column of 21.1 mm, he will try to retune the flute by sawing one-fifth of 25.1 mm (= 5.02 mm) from the headjoint. This is done with the thought that all the lengths will be reduced by this amount and the pitch raised the proper amount.

a. Find the pitch change in cents produced for the notes C_4 to C_5 when the tube is shortened in the manner described.

b. The low-register scale of the flute extends from C_4 to C_5, after which the next octave of a C-scale is continued upwards by a repeat of the same fingerings and a slightly altered style of blowing. Describe the overall trend of pitch error if the modified flute is simply played on its own terms, with no attempt on the part of the player to correct anything by embouchure or blowing adjustments.

c. It turns out that if the flute as a tone-producing machine is very good, the player will have a very hard time using it for serious music with other people. If, on the other hand, it was originally a mediocre instrument, a perfectly acceptable level of musical performance can be achieved by the same player! Explain this apparent peculiarity of behavior.

21-7. Consider the pressure-response curves sketched below for the same woodwind in two conditions: one with a dry and perhaps somewhat porous bore, and the other with a freshly-oiled bore. The difference is exaggerated for clarity.

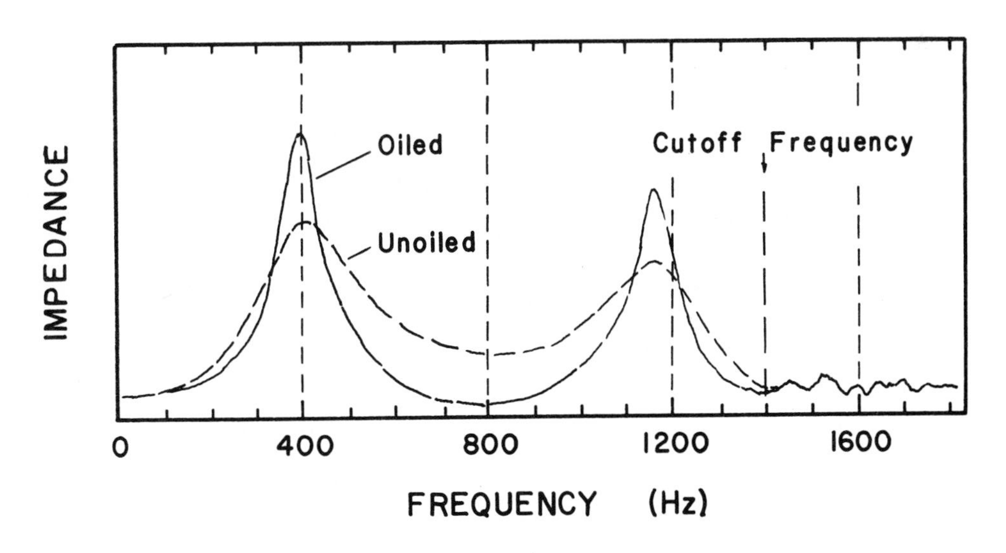

a. Decide whether these curves belong to a clarinet or a saxophone.

b. What concert-pitch note does it play in the low register? What would it like to play in the second register? (Be as quantitative as possible and remember that everything is exaggerated here.)

c. For pianissimo playing it is possible to confine your attention to the frequency region near 400 Hz. Will the oiled or the unoiled version be easier to play? (Give your acoustical reasoning.)

d. For mezzo forte playing one must also take into account the nature of the response curve in the neighborhood of 1200 Hz. As one plays a crescendo with no embouchure modifications, what will you expect the pitch to do? Compare your predictions for the oiled and unoiled instruments.

e. Go back now and look at the frequency region near 800 Hz. In general terms discuss the things that control "ease of playing" (help or hinder) when one uses the two versions of the air column when playing at a mezzo forte level.

f. Musicians like to play on instruments that have good cooperation between the air column and the various components of the produced tone. The improvement in tone and response is such that most players will prefer to use a hard-to-blow instrument with good cooperation to an easy-to-blow instrument with poor cooperation. Explain your reasons for predicting the preferences of musicians for the two conditions of the instrument described above.

21-8. The input impedance curve of a saxophone or oboe-like woodwind has a general appearance as sketched below for a note such as written G_4 played with about half of the tone holes closed.

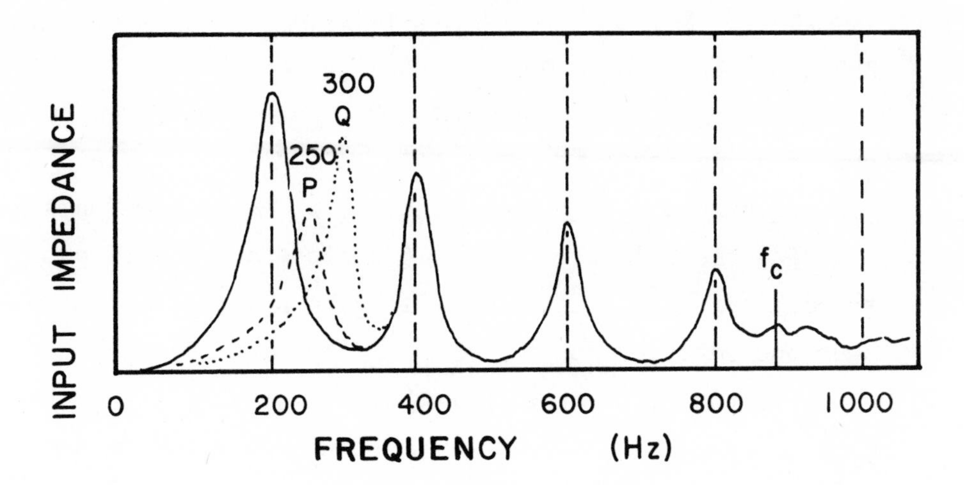

The little resonance peak marked P represents the position to which the lowest frequency resonance is to be moved by the action of the normal register key, while the peak marked Q is a possible position for this peak if a different kind of register hole is designed and installed.

a. Explain in a sentence or two why the peak marked P must be arranged to be quite a lot shorter than its ancestor, and state the upper limit of acceptability for its tallness. Now explain why the optimum value for the frequency of peak P lies at 250 Hz in the example selected for discussion.

b. Explain in musical terms what will happen if the register key for Q is pressed while the player is sounding a low-register note. Does anything change if he is playing loudly or softly other than that which is familiar for all woodwinds? Comment briefly on the possibility of making a useful instrument that is based on this perfectly possible type of arrangement in which both kinds of register keys were supplied on the instrument!

21-9. Refer to the Werckmeister temperament and equal temperament tuning errors in Table 16.1, page 309, of Fundamentals of Musical Acoustics by A. H. Benade. Describe the intonation problems of a performer on a Baroque flute accompanied by a keyboard instrument tuned in each of these ways. The flute has some essential quirks which you must consider. Relative to the notes G_4 and G_5 the other notes (specified in equal temperament) are disposed so that the $F^{\#}$'s are very flat (15 to 20 cents), or on an alternate fingering the $F^{\#}$'s are sharp (15 cents), and the F naturals are both quite sharp (15 cents for the upper one with the lower one worse). Assume that all the other notes are close enough to the proper position that when played in the keys of one or two sharps or flats the flute does not cause conscious uneasiness about the tuning (major keys). All of the notes can be played in tune (minimum beat) on the flute to either sort of keyboard, but the farther one has to reach to get a note the more likely one gets tired, confused, or exhausts his talents on tuning and has none left to expend on the rest of the music. Decide whether the player will be more comfortable with Werckmeister III or equal temperament if asked to play in D major, B minor, and F major.

21-10. Refer to the resonance curve of the horn discussed in problem 20-5, where it was played with a pressure-controlled reed. Extend the discussion to the situation where the horn with this resonance curve is played with a velocity controlled valve as in fipple-type instruments.

a. Explain why tones could be "happily" generated at about 175 Hz, 250 Hz, and 350 Hz. In other words, describe the three regimes of oscillation which would lead to tones whose fundamental component frequencies are the ones given.

b. Which one of these regimes of oscillation would be most stable in pitch if one tried to move them up or down? Explain briefly.

21-11. The adjustment of wind instruments is conveniently done with reference to "perturbation curves" for help in showing the raising or lowering of the various modes produced by changes in the cross sectional area of the air column. Refer to the effects of stiffness and loading changes discussed in Chapter 9 and the problems in section 9.6 of Fundamentals of Musical Acoustics by A. H. Benade, and Chapter 9 herein. Also refer to Chapter 22 of Fundamentals of Musical Acoustics where Figures 22.3, 22.4, and 22.13 show the use of "perturbation curves."

a. Verify that the right-hand set of curves in the diagram below are indeed in the correct form for the perturbation curves to be used for the addition of mass to a freely vibrating string with no attendant change in stiffness. Make the explanation for guitar-string experiments given in part 2 of section 9.6 of Fundamentals of Musical Acoustics a basis for your discussion.

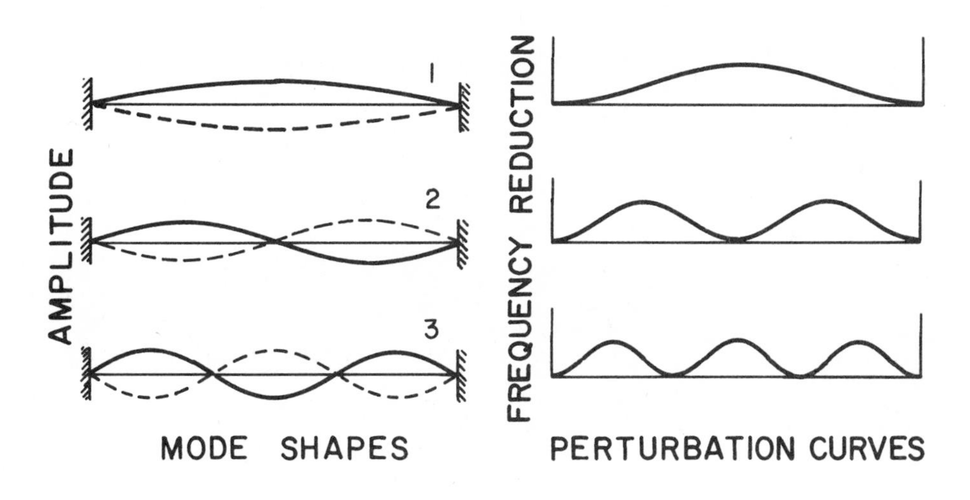

b. Sketch the perturbation curves that would be appropriate if the string were thickened or thinned so that stiffness changes would also need to be considered.

21-12. While the properties of rooms are such as to smooth-out and average-over the differences in the strengths of partials radiated in different directions by an instrument, the details of this radiation are perceptually significant. This is especially true when microphones, recording machines, playback, and loudspeakers are interposed in the acoustic chain between source and listener. The radiation patterns of wind instruments are different for the three situations where the partials are below the cutoff, just above the cutoff, or well above the cutoff. Consider the location of the instrument in the room and the possible reflections from walls or the floor. Where do you put a microphone to get a sensible recording of a clarinet with $f_c \simeq 1500$ Hz if the music runs from E_3 to D_6? Where do you put a microphone for a trumpet whose $f_c \simeq 1400$ Hz?

22

Instruments of the Violin Family

22-1. Consider the output sound a violin bowed mezzo forte at a point one-eighth of the way along the string from the bridge.

a. Would you expect the sound output spectrum to be the same if the bowing were to take place one-eighth of the way along the string from the nut? Explain. It may be helpful to sketch the relationship between the location of the bowing point and the amplitudes of the various modes. Discuss the symmetries that can be used in the explanation.

b. List several reasons why you would not expect the 8th, 16th, 24th, etc. harmonic components in the emitted tone to be missing from the sound.

c. Arrange to observe the bowing of a violin-family string instrument for the conditions stated in part a. Report the results of the observations in a commentary on the nature of the sound for each case.

22-2. For ordinary playing the violin string is bowed in a regime of oscillation that is based on the first-mode response peak. Describe the cooperation of peaks which constitute the regime of an oscillation called "playing on the third natural harmonic" of the string. What might happen if the performer tried to play this while bowing one-third of the way along the string?

22-3. Consider a viola whose main air and wood resonances are located at 230 Hz and at 360 Hz.

a. Make a list of the named notes in the playing range of the instrument for which one or more of the partials are especially influenced by the presence of these resonances.

b. On the basis of part a suggest musical key signatures that might carry a particularly strongly-marked flavor as a result of the unique resonances for this instrument. How do these implications correlate with the normal behavior of violas in regard to tone color?

22-4. The main air and body resonances of a guitar play a role in its sound output that is exactly analogous to the corresponding resonances of a violin.

a. Devise an experiment to find the air resonance on a guitar. (It generally lies about three semitones above the bottom note of a guitar at about 95 Hz.)

b. Continue the experiments to find the first three mode frequencies of the top plate. (See Figure 9.7 of Fundamentals of Musical Acoustics by A. H. Benade; the values will be in the neighborhood of 185, 285, and 450 Hz.)

c. Describe the musical implications of this set of resonances (identifying any special notes, influences on key signatures, etc.) and relate them to any knowledge you have of your guitar.

22-5. In the loudspeaker systems termed a ducted port or bass reflex the natural frequency of the speaker cone is a cognate of the main wood resonance of a violin or guitar top plate. The lowest mode for the air in the cabinet is the cognate of the main air resonance of the body cavity of the instrument. Consult Figures 24.3 and 24.4 of Fundamentals of Musical Acoustics by A. H. Benade for reference to a description of suitable criteria for proportioning the resonance frequency and damping for this frequency of the cabinet so it works best with a loudspeaker whose resonance is at 40 Hz. You will have to specify clearly in acoustical terms your meaning of the word "best" for the loudspeaker system.

22-6. a. Suppose two violins are playing different tones a musical third apart. Compare the perceptual "noisiness" of any inherent tuning errors associated with the intervals just major third, 5/4; just minor third, 6/5; pythagorean third, 81/64; and equal temperament third. Assume the errors in tuning are to be influenced by the 20 cent bandwidth of each component of the bowed string in a way that is analogous to notes played on the triples of real piano strings. (See Chapter 17, Fundamentals of Musical Acoustics by A. H. Benade.)

b. Contrast the conclusions formed about noisiness in part a with the noisiness of analgous intervals produced by two woodwinds, brasses, or voices.

23

HALF-VALVED OCTAVES, BURRS, MULTIPHONICS, AND WOLF NOTES

23-1. The wolf-note behavior described for bowed string instruments is a very close cousin to the pulsating sound an imperfect saxophone will make at the end of an extended diminuendo in the low register.

a. Translate the wolf-note discussion for violin-family instruments so that it applies to a saxophone air column. For the instrument under consideration there are four active resonance peaks below cutoff: 155, 302, 451, and 600 Hz.

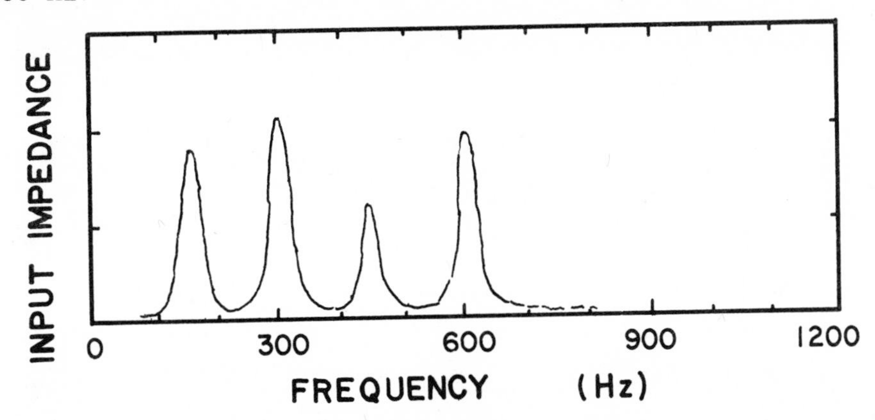

b. Apply similar arguments to a clarinet-like air column for which there are resonance peaks at 103, 300, 500, and 700 Hz. Recall that there are minima in the resonance curve at 200, 400, and 600 Hz. What assumption should be made for the relative tallness of the peaks?

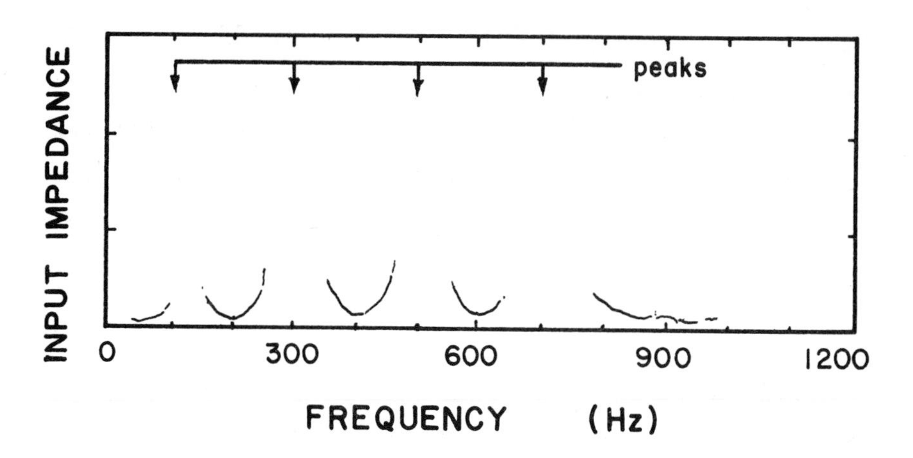

c. What can be done to alter the air columns to correct the error that produces the resonances given in a and b? You may assume that you have no responsibility in this exercise for correcting or preserving the behavior of any other notes.

23-2. The "pedal note" of a trumpet or trombone is sustained by air column peaks 2, 3, 4, 5, and 6 that feed the similarly numbered harmonics of the played tone. The pitch of this "pedal note" is an octave below that of a normal played with the cooperation of peaks 2, 4, 6, and 8. The measured sound spectrum of a "pedal note" contains a significant amount of the fundamental component. This occurs despite the fact that this component lies in the valley between peaks one and two of the measured resonance curve (Figure 20.6 in Fundamentals of Musical Acoustics by A. H. Benade). Explain the origin of this very real member of the measured sound spectrum. Note that the heterodyning of the ear as it relates to pitch perception does not require the presence of this component.

23-3. Since the time of Beethoven at least, horn players and trombonists have been able to fill in the gap between the "pedal note" and the first regular note of the open horn sequence. The "pedal note" is obtained for the horn by the method discussed in problem 23-2. The fill-in note is obtained by the use of a tone played a musical fifth above the "pedal note." Deduce the nature of the regime of oscillation belonging to this note. Your description should include a comment on the crucial role required for the third air column resonance peak.

23-4. Consider the resonance curve belonging to a much-simplified multiphonic fingering on a woodwind.

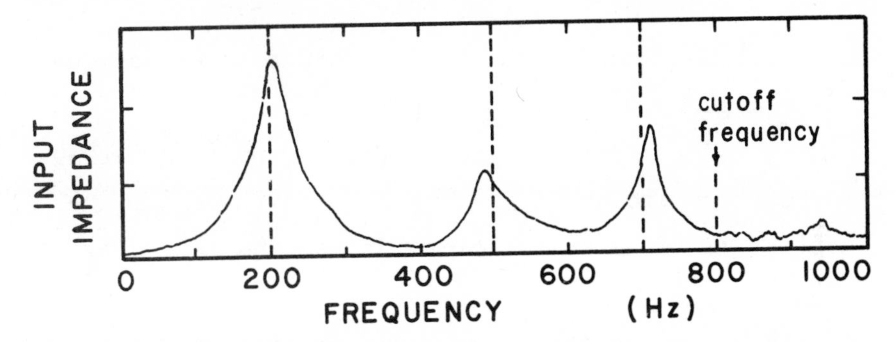

Begin by explaining why an ordinary regime of oscillation at mezzo forte level would not be possible on this air column. Then describe a kind of regime of oscillation that gets its energy input via partials having frequencies near 200, 500, and 700 Hz, which share their influence on other peaks in the regime through heterodyne action in the normal way. What other components will be present in the spectrum? What set of pitches would you expect a listener to assign to the complete collection? (Note: There are a number of ways in which this part can come out depending on your assumptions, so be consistent as you develop your explanation.)

23-5. An individual with your experience in musical acoustics should be able to analyze an acoustical situation or a musical instrument and apply the methods of acoustics to it. Pick some topic having to do with tone color, intonation, voicing, etc. of an instrument or the voice, construct a question, and then supply the answer. Alternatively, you can formulate a question about listening situations and then supply the answer. As a help in formulating your own question some ideas are given here.

a. Why are the hammers at the bass end softer and larger than those at the treble end of the piano? Why do they not vary in hardness or width by as many factors of two as there are octaves on the keyboard?

b. Why do you suppose pianos tend to have several strings on each note? Loudness is hardly a sufficient reason. Why? What else might be happening?

c. What can be proposed as the reason for mounting several pickups on electric guitars?

d. Why are the sounds different for the various methods used to amplify acoustical instruments, such as contact microphones, accelerometers, and recording microphones?

e. Suppose one plays a harpsichord with a piano technique rather than a proper harpsichord technique. Will the sound have different tone color? Why?

f. What acoustical principles are used when a mute is added to a brass instrument?

g. What is the effect of a mute on a bowed-string instrument?

h. Discuss tone production and color in a reed organ.

i. Compare various fingering techniques for baroque flutes or recorders.

23-6. As an example of the effect of adding a small mass to a vibrating object in order to perturb its motion, consider again the string in problem 8-2. Consider the situation where the violin A-string is modified by loading it with a small mass of less than 1/100th the mass of the string. The mass is located about three-eighths of the way along the string. Were one to bow such a string, the sounds produced would indeed be unusual and very hard to control.

The modes are modified by the mass and depart from the usual harmonics. It is possible to show that the frequency of the string is moved by a small amount approximately equal to the fractional change in mass. Assume the fractional change in mass is 1/100 and that the load is placed exactly three-eighths the way along the string. Find the approximal frequency for the first few modes of the string. Helpful graphical data are found in problem 21-11. Use the curves to estimate the effect of the small mass on each of the three modes. For those students who like formulas and have a hand calculator, the fractional change in frequency of the nth mode of a loaded string is given by the equation

$$\frac{\Delta f_n}{f_n} = \frac{m}{M} \sin^2\left(180 \frac{nx}{L}\right)$$

where m is the added mass, M the mass of the vibrating string, f_n the frequency of vibration of the string, L the length of the string, and x the location of m from one end of the string.

24

SIMPLE VIBRATING SYSTEMS

24-1. Assume that the speed sound travels in air is 345 meters/sec or 1130 ft/sec. Find the time required for a sound to travel one-fourth of a mile, and one-half a kilometer. For additional practice in calculating the relationship between distance, speed, and time consider this version of the problem. Assume that you wish to estimate the distance between you and a lightning stroke by using the time interval between seeing the flash and hearing the thunder. How far does a sound pulse travel in 1 sec?

24-2. For practice in computing pressure in terms of the force per unit area you may wish to consider a somewhat realistic problem.

a. Suppose a stereo stylus has an area of contact with the plastic surface equivalent to a circle of radius 0.0254 mm. The tone-arm is set for a tracking force equivalent to that produced by 1.2 grams. What is the resulting pressure? Why is "a force of 1.2 grams" an incorrect term?

b. Next suppose the stylus encountered a dust particle and the radius of the equivalent contact area now became 2.54×10^{-4} mm. What is the pressure for this case? Comment on the implications of this calculation on record wear and scatches produced by stylus and tone-arm travel on the record were the stylus to skate across the record.

24-3. Consider any vibrating thing that displays the property of elasticity. Suppose that a test is conducted to find its stiffness, and the data in Table 24.1 are obtained for deflection as a function of added mass. See, for

Table 24.1

Deflection as a Function of Added Mass

Deflection (mm)	Mass (gm)
1.0	10
2.0	20
3.0	30
3.5	40
4.0	50
4.3	60

example, The Acoustical Foundations of Music by J. Backus, page 107, or The Fundamentals of Musical Acoustics by A. H. Benade, page 263.

a. Make a sketch of deflection as a function of added mass. Re-label the mass axis in force units but don't re-draw the curve.

b. Consider the linear portion that is described by the relationship F = sx where s is a constant. What is the constant s? For certain vibrating systems the displacement from equilibrium is arranged to occur in this linear region.

c. However, in a number of situations the vibrations occur where the stiffness is not a constant. For the hypothetical example in Table 24.1 of the behavior of vibrating systems estimate the change in stiffness with increasing force. Did the stiffness increase or decrease in this example?

24-4. a. For practice draw a sinusoid that represents simple harmonic motion, and then sketch another curve on the same graph representing a motion that differs in initial phase by one-third, one-fourth, and three-fourths period from the first.

b. Cite some examples from your own experience of simple harmonic motion. Explain the situation where a freely oscillating system may be perfectly periodic with a constant repetition rate and the motion is not a simple sinusoid by sketching the graph representing its motion.

24-5. In the physics of vibrating systems the concepts of work, energy, and the effect of forces are considered in some detail. Discuss these concepts as they apply first to a simple process and second to a vibrating system, using the following imagined experiments.

a. Imagine hammering two boards together with a nail. The nail bends when it is halfway through the second board and must be removed. Does the hammer do physical work on the nail when it is driven into the wood? Does the nail do physical work on the wood? Identify the forces during the driving phase of the process. The claws of the hammer are used to pull the nail. Is work done by the hammer on the nail? Is work done by the wood on the nail? When the nail is rapidly withdrawn and touched it is observed to be hot. What caused the nail to become hot? Would you suppose the nail was heated during the driving operation? What total energy had to be supplied by the person to drive and remove the nail? Would it make any difference if there were only one board into which a nail was driven and then removed?

b. Imagine a taut string set into motion by striking it with a small mallet or hammer. Does the hammer do work on the string? Does the string do work on the hammer? Was energy imparted to the string and in what form? As the string continues to vibrate the amplitude decreases until the motion ceases. What happened to the energy of the string? Include the fact that the string has end supports. (Note: Both parts of this problem can become very complicated very quickly if extended to fine details. Confine your comments to the obvious simple physics aspects. I caution you again to resist drawing implications of a musical sort about hammering a string to set it vibrating at this point. The problems in Chapters 8 and 17 treat some aspects of hammered strings.)

24-6. Consider the decay of vibration of a particular object that shows a single frequency of repetition. The graph of the amplitude of the end of one tine is sketched below.

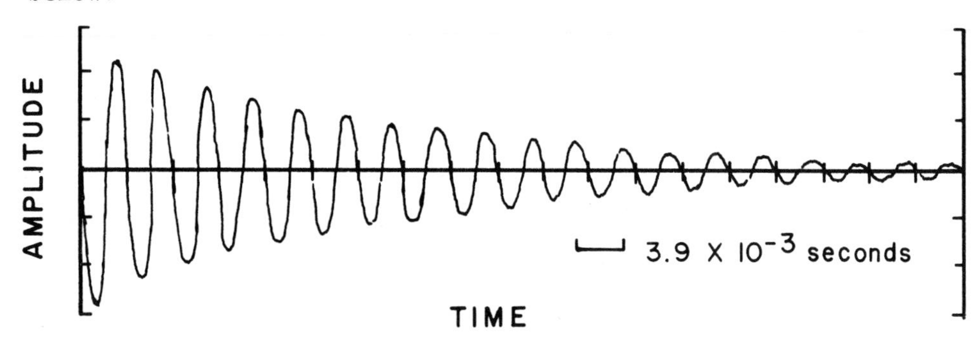

What is the frequency of the oscillations? What is the time for the amplitude of vibration to decay by one-half?

24-7. Define the concept of uniform motion in which you include motion in a straight line and in a circle. From your own experience identify and discuss an example of each type of uniform motion. Discuss an example that shows the relationship between uniform circular motion and simple harmonic motion. (Note: This is a very popular physics lecture demonstration.)

24-8. A mass and spring can form a simple vibrating system. What properties of the system determine the frequency of vibration? Can the system of a single mass and spring show more than one characteristic way of vibrating? Before you conclude your work on this problem consult the library and consider the spring and mass system described by Reverend Wilberforce. Perhaps your physics department has a Wilberforce pendulum in its collection of lecture equipment. Arrange to observe its behavior as an example of vibrating systems. It is an interesting illustration of some physics principles.

24-9. Construct your own vibrating system from rubber bands, for the restoring force, and some unbreakable object such as a washer or ring. Set the system into motion in the spring and mass manner (not as a pendulum) and adjust things so you can time the vibrations. Find the period of the system and the time for the amplitude to decay by one-half. You could estimate the amplitude by viewing the motion beyond the edge of a vertical ruler. Double and triple the number of identical objects suspended from a single rubber band and report the period and halving time. Repeat for two rubber bands and the same collection of objects. What is the stiffness of a single rubber band? Rank the frequencies you found from the lowest to the highest. What is the ratio of each frequency to the lowest frequency?

25

Wave Propagation

25-1. Consider a single frequency vibration of the mass-spring type. Construct a graph of the displacement from equilibrium, velocity, and acceleration of the mass. Assume that the displacement graph is a sinusoidal function of the cosine type so the initial displacement is positive and above the equilibrium position. Just sketch the curve and identify some characteristic positions of the mass and the corresponding points on the graph. Do the same sort of identification for the separate graphs of velocity and acceleration.

25-2. Assume that a sound wave of single frequency travels to the right in air and in your diagram. Sketch the curves that represent the pressure amplitude and displacement amplitude. Use the concepts of simple harmonic motion to discuss in your own words the fact that the extreme amplitudes do not occur at the same time.

25-3. What is the difference between the speed of sound and the average speed of a "particle of the medium"?

25-4. Reflection of sound gives rise to the phenomena of echoes. Problems that involve the time for echoes or between echoes require the speed of sound and the distances of the reflecting surfaces.

a. The question "if an echo is heard 0.35 sec after a sound is made, how far away is the reflecting surface" is incomplete as it stands. What would you assume for the speed of sound on a mid-summer day? On a winter day? How far away would the reflecting surface be for the situation where your assumed or calculated speeds apply?

b. In a place where the speed of sound is 345 meters/sec, a wall 20 meters away returns a sound from it in what time? You probably assumed, without stating it, that the source and observer were 20 meters in front of the wall. What would one observe if the sound were a short impulse and formed at a location 20 meters further away from the wall than the observer in a direct line?

25-5. Discuss the reason a person can hear the echo from a row of closely spaced, narrow, tall trees or a fog bank.

25-6. The refraction of sound is an acoustical phenomenon usually encountered outdoors.

a. Refraction can account for the fact that one can hear better at greater distances (sound strength is greater for a constant strength source) on a cold crisp autumn or winter morning. Discuss the conditions that are necessary for this situation.

b. Explain why it is easier to hear a sound source located at ground level on the down-wind side. Assume that the wind is blowing at a steady rate and that the wind speed increases with height above the ground. This is a realistic requirement for the behavior of a moderate to heavy wind. I find it interesting to first draw a series of concentric circles with a common center of ground level. Then graphically draw horizontal lines starting at points on each circle that represent the distance of travel with the wind speed at that height. When you do this you need to attend to such things as addition of vectors that represent distances travelled in the appropriate times. If you do this carefully the conditions for refraction of sound by the wind can be described. (Note: You must have a change in wind speed with height for this refraction to occur. Why?)

c. Loud noises such as explosions or cannon-fire can be heard at extremely great distances. Reports exist of sounds of this sort travelling hundreds of miles. It is not always clear in the accounts that the reporters or commentators include all the possible contributions to the phenomenon. Consider typical weather conditions and speculate on ways that refraction and reflection could contribute to the long-distance travel of a sound.

25-7. The diffraction of sound can explain how one hears around corners, behind posts, and at the side of openings in walls. Consider two situations that illustrate the phenomena.

a. One speaks from a distance of 2 meters behind a window that is 60 cm high and 20 cm wide. The sounds of the voice are imagined to range roughly from about 15 Hz to 3500 Hz for this situation. What effects would diffraction have on the propagation of sound through the window? Give a rough estimate of these effects in a comment on the direction and strength changes for the diffracted sound. What would be your answer if the long direction of the window were horizontal?

b. A plane sound wave is incident upon a wall with a small hole in it. Describe the directional characteristics of the sound that emerges from the hole. Include in the discussion cases where the wavelength of the sound is shorter than, the same size as, and longer than the diameter of the opening.

25-8. When two sounds travel through the same point in the medium, the combined effect can be described as the combination or superposition of the individual effects.

a. Draw a curve representing a wave of a single frequency so that at least three complete repetitions of the wave are shown. Second, on the same axes draw a curve of a different single frequency that represents a disturbance at the same location in the medium. The amplitude of this second curve is one-half that of the first, and the frequency is twice that of the first. On the same set of coordinate axes draw the combination of the waves.

b. On a single set of coordinate axes draw three single frequency curves representing vibrations of 60 Hz, 120 Hz, and 150 Hz. The curve should show at least four complete repetitions of the 60 Hz wave form. The three curves should have the same amplitude. Draw the sum of these three curves. What is the repetition rate of the resulting summation curve?

25-9. Suppose you are outdoors listening to the sounds radiated by a pair of small loudspeakers. In this example the signal to each loudspeaker is from the same electrically generated sinusoidal current.

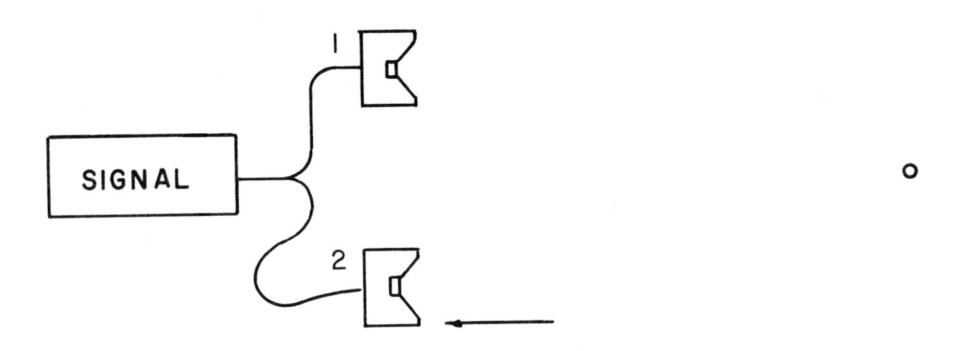

If one of the loudspeakers (signal number 1) is kept in a fixed location, and the other (signal number 2) is slowly moved farther away from the ear, consider the strength of the combined signals at the ear. (Note: In b and c the sounds are sufficiently intense at the source to be audible at the ear so that one could observe the results.)

a. Under what conditions of distance from the loudspeakers will the combined sound be most intense?

b. Under what conditions of distance from the loudspeakers will the combined sound be a minimum?

c. Can there be a situation of distance from the sources in this outdoor setting where the combined sound will be faintly audible, or nearly so?

26

Idealized Systems, Strings and Air Columns

26-1. In order to demonstrate the nature of the vibration of a tuning fork in a classroom the fork is struck a proper but rather forceful blow. Its audibility while held in the air is rather low. When it is touched to a tabletop or small box the increase in audibility is dramatic and the sound it makes is easily perceived in a large room. Why is the fork more easily heard when held against the object? Compare the audible duration and also the physical halving time when the fork is held in the air in comparison to being held against the tabletop. Why are these not necessarily proportional?

26-2. A tuning fork of unknown frequency is sounded with a standard 880 Hz tuning fork and found to give rise to four beats per second. In order to determine the true value of the unknown frequency a wad of chewing gum is placed on one of the prongs of the tuning fork of unknown frequency and it is observed that the beats are now six per second. What was the unknown frequency? Carefully explain the reasoning behind your answer.

26-3. One can consider a resonance to represent a frequency selection property of the system. What are more specific ways to specify this property other than broad or narrow? You may wish to refer to Figure 10.9, page 157, of <u>Fundamentals of Musical Acoustics</u> by A. H. Benade; and Figure 4-12, page 99, of <u>Vibrations and Waves</u> by A. P. French for some helpful ideas; or Figure 18, page 74, of the Acoustical Foundations of Music by J. Backus.

26-4. First sketch the modes of a flexible string of uniform diameter that is fastened at each end to a rigid support.

a. The frequency of the lowest mode is 256 Hz and the length of the string is 1 meter. Find the speed of propagation of the vibrational disturbance in the string.

b. If the tension in the wire was 5 newtons, what would be the mass of the 1-meter long wire?

c. Suppose one tapped the string at one end in a way to give it a small transverse impulse of displacement. How long would it take the pulse to return to the place where it was tapped?

d. What is the period of vibration of the lowest mode?

26-5. Sketch the pressure patterns of the first four modes of a long cylindrical air column of uniform diameter that has both ends terminating in an infinitely stiff closed end. Indicate the regions of large and slight pressure variations. If the column is 65 cm long and the speed of sound in air is 345 meters/sec, what is the lowest frequency of oscillation of the air in the tube?

26-6. Sketch the pressure patterns of the first four modes of a long cylindrical air column that has one end open and one end terminating in an infinitely stiff closed end. Since the pressure drop cannot be exactly zero at the open end, describe the location of the implied pressure node.

a. If the column is 65 cm long and the speed of sound in air is 345 meters/sec, what is the lowest frequency of oscillation of the air in the tube? Usually one finds this problem worked without the end correction. Use an approximate expression for the end correction and find the difference between the two results.

b. Discuss some physical reasons for the end correction, including some possible reasons for a dependence of the correction on frequency.

26-7. The pressure patterns of the modes of a long cylindrical air column that is open at both ends is said to be similar to the amplitude patterns of a string fixed at both ends.

a. Compare the mathematical equations used to describe the frequencies of the various modes and discuss the similarities and differences.

b. If one extends the discussion from the highly simplified view to a more realistic view of the physical properties of the ends of each object, some similarities are apparent. Are the end corrections actually similar? Discuss briefly the complications that result. The main idea of this consideration is to emphasize the point that the idealized strings and air columns are convenient for mathematical calculations of some properties of vibrating systems. Such systems can form a starting point for consideration of real systems if one is willing to accept these limitations and consider each application in detail.

26-8. Find the "half-wavelengths" in air for sounds of the following frequencies:

30, 220, 440, 635, 3000, 12,000 Hz

Assume the temperature is 20° C.

26-9. Most real oscillating systems of the type called free oscillators exhibit a slow decay in amplitude of oscillation. The decay can be related to the resistance (or impedance) to motion of the system. The loss may be internal to the system such as the internal frictional behavior of a metal wire. The loss may be frictional as the object attempts to transfer its vibrations to the supports and surrounding air. Select a simple oscillator such as a vibrating string, air column, jug, pendulum, and discuss the possible causes of the damping and the effect on the oscillations.

26-10. Discuss the essential characteristics of a cavity system of the type known as a Helmholtz resonator.

a. Compare the frequencies of two cavities with the same volume and different diameter openings, but having the same effective length.

b. Compare the frequencies of two cavities with the same opening and different volumes. Test this case with two pop bottles of different volume.

c. Compare the frequencies of two cavities where the volume and diameter are constant but the length of the opening changes. Test this case by making several rolled paper tubes of different lengths and inserting them (so they fit snugly) into the neck of a bottle.

26-11. Obtain a ball point pen and unscrew the barrel of the pen so you can remove the cartridge. An inexpensive plastic style will do very well. Close the small end of the pen barrel with a finger and blow across the larger screw-threaded end to obtain a sound. The technique is similar to obtaining sound from a gallon jug or pop bottle. Estimate the frequency of the sound by comparison with a standard. For the estimate a piano and Table 2.1 as a reference are handy. Next reverse the pen barrel by closing the large end and blowing across the small end. This may require a bit of practice. Try to overblow either or both ends to produce higher pitched sounds. Discuss the behavior of the resonance of the pen barrel in terms of cylindrical air columns and Helmholtz resonators. Verify that neither model is exactly correct by using the frequencies you discovered and the equations for the two models along with the measurements of the pen barrel.

26-12. Obtain a plastic bottle with thin sides that can be used as a "Helmholtz" resonator. Hold the bottle lightly or rest it on a surface and produce a tone with your best technique. Next, hold the bottle with your hands to reduce vibrations of the side wall. Discuss the change in pitch that results. (You may find it advisable to experiment with bottles of various shapes.)

Glossary of Terms used in this Problems Book

Absorption—In acoustics, absorption refers to that portion of the energy in a sound wave that is neither reflected nor transmitted at a surface. The acoustical energy is converted to some other form, usually heat, in the material.

Acoustics—The behavior of sound, treated in all its aspects. Commonly, the term is restricted to a study of the transmission of sound through various media or in various enclosures or conduits, including the effects of sound reflection, refraction, interference, diffraction, and absorption. Often, the term includes the perception of sound, including the working of the ear and the processing of signals from the ear by the brain. These usages of the term acoustics are often identified by the term physical acoustics for the former and psychoacoustics for the latter.

Amplitude—The greatest magnitude of an alternating quantity in either the positive or negative direction.

Antinode—The position of maximum amplitude in the static or steady pattern of vibratory disturbance which lies between two adjacent places of little or no disturbance.

Aperiodic—A vibration which does not repeat itself, hence a non-periodic one.

Audibility—The ease with which a sound may be heard.

Audiogram—A chart showing the hearing response of the person as it depends upon the frequency of vibration of sinusoidal (single frequency) sound.

Band—A segment or part of the frequency spectrum, response curve, or impedance curve.

Bandwidth—The band of frequencies over which the spectral response falls to within specified limits of the maximum value, usually one-half.

Beat—A sound which has a periodic rise and fall in amplitude and heard as a periodic change in loudness. Beats are caused by combined effect of the sound from two sources of slightly different frequencies.

Bridge—In stringed instruments (violin, etc.), the wooden support atop the table across which the strings are stretched or to which the strings are attached. It serves to raise the string and transfer vibrations to the instrument proper.

Buff stop—An arrangement in a harpsichord where a soft piece of buff leather is pressed against the vibrating portion of each string at the nut.

Cent—The unit for the scientific measurement of musical intervals. It is one hundredth of a half tone (semitone); thus the half tone equals 100 cents, and the octave contains 1200 cents.

Cutoff frequency—A characteristic frequency often designated f_c at which the spectral response falls over a narrow frequency range to a small or zero value. The term upper cutoff frequency refers to the case where the small response occurs above f_c and lower cutoff frequency refers to the case where the small response occurs below f_c.

Cycle—In single frequency sinusoidal motion, one complete repetition from a starting point, through both extremities, and back to the next equivalent point.

Damping—The amplitude of a freely vibrating system with frictional losses will diminish with time until it approaches zero. Damping is the result of energy loss in the system that may occur as a resistance to motion.

Diffraction—Occurs when waves bend around an obstacle in order to move into the space behind it.

Echo—When a sound pulse is incident upon a surface of large area some part of the sound energy is reflected. If the time interval between the emission of sound and the return of the reflected wave is sufficiently long, the reflected sound heard after a silent interval is called an echo.

Embouchure—The position of the lips in the playing of wind instruments.

Excitation—The process by which a force acting upon an object produces a response. A repetitive force can induce a vibratory response in an object in the excitation process.

Exponential—A mathematical function that increases or decreases in magnitude as obtained by raising the base of the natural logarithms (denoted e with a value of 2.718 ...) to some positive or negative power. Any quantities that vary or change according to the exponential mathematical function.

Fipple—Generic designation for instruments of the recorder or flageolet type. Fipples are also found in ordinary whistles or ocarinas.

Flutter—Rapid changes of frequency.

Flutter echo—A rapid succession of reflected echoes resulting from a single initial sound.

Force—The agent that alters or tends to change the state of rest or motion of an object.

Formant—The peaks associated with the spectrum envelope of the sounds produced in vocalization or singing. The vocal folds produce a spectrum of sound with many harmonics that is attenuated upon passage through the vocal tract. The resonance frequencies of the vocal tract can be related to the spectrum envelope peaks. The vocal tract has four or five important formants. The term formant can be used to identify the similar occurrence of peaks in the spectrum envelope of musical instruments.

Frequency—The number of complete repetitions, oscillations, or cycles in unit time of a vibrating system. Formerly stated in vibrations per second, cycles per second, but now in hertz, Hz.

Fundamental—The primary or parent frequency of a set of harmonic partials, all with frequencies that are integer multiples of it.

 Halving time—The time required for the amplitude of an oscillatory, repetitive, or vibratory motion to decay to one-half amplitude.

Harmonic—A sinusoidal component of vibration that is an integral multiple of a lowest frequency. The lowest such frequency is called the fundamental.

Hearing—The process by which vibratory stimuli are processed by the ear into the perception called sound.

Heterodyne—The process by which a system responds to the vibratory stimuli with an output that has new frequencies that are multiples, sums, and differences of the input frequencies. Such a response is characteristic of a nonlinear system, including hearing.

Humps—The name given to locations of maximum disturbance in the static or steady pattern of vibration of an object.

Impedance—The degree that an object resists the action of an external force or pressure. The ratio of applied force to rate of displacement of the object. The ratio of the alternating sound pressure to the rate of volume displacement of the surface that is vibrating. The degree to which sound is reflected or transmitted at a boundary depends on the ratio of wave impedances of the two media.

Impulse sound—A sound of very short duration.

Inharmonicity—The extent by which the frequency of a partial differs from an integral value of a fundamental.

Interval—The perceived distance or difference between two pitches.

Intonation—Degree of adherence to correct pitch. Good intonation implies close approximation of the pitch; poor intonation implies deviation from pitch. Correct pitch is governed by musical taste and tradition.

Key note—Starting note or tonic of a musical scale.

Kilogram—A unit of mass defined for a particular cylinder of platinum kept at the International Bureau of Weights and Measures. It is approximately the mass of 1 liter of water.

Kinetic energy—The energy of motion possessed by a moving object.

Logarithm—The logarithm of a given number is the exponent or power to which another number, the base, must be raised to equal the given number. Two bases are widely used. The natural logarithms have 2.71828 ... as the base, denoted by the symbol "e." The common logarithms have 10 as their base.

Masking—The effect by which one sound or component of a sound influences or hinders the perception of another.

Mode—Any one of the static or steady patterns of disturbance of a vibrating system or of the oscillatory motion of an elastic object.

Node—A point or line of no or very small disturbance in the static pattern of disturbance of a vibrating system.

Noise—Sound unwanted by the listener, meaningless sound, random sound, an aperiodic sound.

Note—A musical sound of specified pitch produced by a musical instrument, voice, etc. The designation of a pitch in written music.

Octave—The interval embracing eight diatonic pitches. Also the eighth pitch of the diatonic scale. Acoustically this is a pitch with twice the frequency of the first. Also, a difference in frequency of a factor of two. (800 Hz is an octave higher in frequency than 400 Hz.)

Oscillation—The repetitive motion of an elastic object. The repetitive variation in force or pressure.

Oscillogram—A record of sound pressure, amplitude, or a representative physical quantity of oscillatory motion as a function of time.

Partial—The individual single frequency sinusoids making up a combination sound or multicomponent sound. The term partial is a more general term than harmonic, which is restricted to an integral series of partials.

Peak—The portion of a graph of excitation or spectral response with a large amplitude in a small range or band of frequencies.

Period—The time for one complete cycle of vibration or repetition of an oscillatory motion.

Periodic—Motion which repeats itself indefinitely.

Perturbation—Any influence that produces a modification (usually small) in the vibratory behavior of a system more specifically as a change in frequency or amplitude.

Phase—A measure of whether a sound or other periodic function is "in step" or "out of step" with a reference. It is measured as an angle either in degrees or in radians (360° or 2π radians) or as a part of a cycle (e.g. a quarter cycle out of phase).

Pitch—The property according to which tones appear to be high or low in the perceptual cognate or correlate of frequency.

Potential energy—A form of stored energy. The energy due to position or mechanical strain.

Pressure—The amount of force acting on each unit area.

Radiation—The process by which energy is transferred or carried away from a source through a medium.

Recipe—The specification or designation of vibrational components of a sound. The regime or list of possible modes of an elastic object or vibratory system. The specification often in graph form of the spectrum of sound pressure values, vibrational amplitudes, or response peaks of a system.

Reflection—When a wave encounters a sudden change in properties of the medium (i.e., wave impedance), the direction of propagation of the wave is changed. The angle of incidence is equal to the angle of reflection, and a greater or lesser amount of the wave is transmitted (refracted) into the second medium.

Refraction—The change in direction of a wave when it moves through a medium of gradually changing properties (wave impedance), or through the boundary of two media of different properties (wave impedance).

Repetitions—In acoustics, repetition refers to the designation of the succession of periodic or identical patterns (waveform) of an oscillatory or vibrating motion.

Response—The extent or process by which a vibratory body moves or changes under the influence of a repetitive applied force or stimulus.

Reverberation—Sound persistence due to repeated boundary reflections after the source of sound has stopped. The aggregate effect of the dying away of the numerous vibrational modes of the room.

Reverberation time—The time it takes for reverberant sound of a given frequency to decay by 60 dB after the source is stopped.

Scale, intensive—Scales by which measured sensations or quantities can be ranked in order, as with increasing magnitude relative to a reference.

Scale, musical—A sequence of tones or pitches containing musical intervals that are useful for musical composition and performance.

Scale, numerical—Scales by which the quantity being measured can be ranked in an order where each unit is a numerical value.

Semitone—An interval of approximately one-twelfth of an octave.

Simple harmonic motion—A vibratory motion of an elastic object that has a single frequency, regularly recurring pattern with a graph of displacement that is a sine or cosine function.

Sinusoid—A single frequency vibrational component of a sound. A sound that has a single frequency of repetition represented in a graph as a sine or cosine function.

Sone—A perceived unit of loudness of a sound for a given listener equal to the loudness of a 1000 hertz sound that has a sound pressure level 40 dB above the listener's own threshold.

Sound—Sound is a sensation produced by vibration. It is the sensation experienced by vibrating air particles interacting with the eardrum. Sound can be perceived if the transmission is by a solid or a liquid. Vibrations of the air, especially very low frequency vibrations, can be felt by other sensors than those in the ear. Normally such a sensation is not termed sound. In a sound the vibration of the air particles is the collective vibration as a pressure variation, not the random motion of the air molecules.

Sound pressure level—Twenty times the logarithm to the base ten of the ratio of the sound pressure divided by the sound pressure corresponding to an arbitrarily chosen reference that lies near the threshold of audibility for a 1000 Hz sinusoid.

Spectrum—The representation in tabular or graphic form of the analysis of a tone into its components or partials.

Speech spectrum—The amplitude, sound energy, or loudness versus frequency of each partial of the speech sound.

Spring—An elastic device, such as a coil of wire, that regains its original shape after being compressed or extended.

Spring constant—A measure of the stiffness of a spring. It is measured in units of force per unit displacement.

Standing wave—A static or steady vibratory disturbance in an elastic body with fixed, stationary places of little or no disturbance.

Stiffness—The ability of a system to resist a deflecting force.

String—A long piece of material of small diameter, completely flexible, and pulled at its two ends by equal and opposite forces that keep it taut.

Temperament—General designation for various systems of tuning or setting pitch in which the intervals are "tempered," or adjusted so they deviate from the intervals obtained from frequencies with integer ratios according to a particular musical taste or tradition.

Tension—The magnitude of the pulling force which keeps a string taut.

Tone—A sound made up of a reasonably small collection of harmonically related components. Its pitch behavior is stable and uniform under many experimental conditions.

Tone, musical—A compact group of sinusoidal harmonic components in a sound of definite pitch.

Tonic—The first and main note of a key or scale.

Transient—A sound or component that decays rapidly.

Triad—A chord consisting of three notes: a fundamental note, its upper third, and its upper fifth. It may also be described as consisting of two superimposed intervals of a third.

Tuning fork—A two-pronged metal object set in vibration to produce a sound which serves to check pitch. Its sound has a single frequency without partials when carefully set into vibration in its tuning mode.

Vibration—The rapid to and fro motion characteristic of an elastic solid or a fluid medium (air, various gases, water, or various liquids) influenced by such a solid.

Vibration, free—A spring system or elastic body which oscillates without any outside influence.

Vibrato—On violins, cellos, etc. a regular, minute fluctuation of pitch. In singing the term usually denotes a slight wavering of pitch.

Vocal folds—Folds of ligaments on each side of the larynx.

Vocal tract—A tube which consists of the pharynx, mouth, and nose. The shape of the vocal tract can be varied extensively by moving the tongue, lips, soft pallet, and larynx.

Voice—The human means of producing sound by the action of the lungs, vocal folds, and vocal tract. The adjustment of an instrument, especially wind instruments, pipe organs, and harpsichords, to fit standards of pitch, tone color, responsiveness, etc.

Wave—A periodic vibratory disturbance which is transmitted in a medium.

Waveform—The pattern of one cycle of a periodic vibration.

Wavelength—The distance from one point in a wave or vibrational disturbance to the corresponding point, as for example the distance from one crest to the next in a water wave or from one pressure maximum to the next in a sound wave.

Whole tone—An interval that is approximately one-sixth of an octave.

INDEX